Drilling Engineering Handbook

Drilling Engineering Handbook

Ellis H. Austin

International Human Resources Development Corporation • Boston
D. Reidel Publishing Company, A Member of the
Kluwer Academic Publishers Group, Dordrecht/Boston/Lancaster

Library of Congress Cataloging in Publication Data

Austin, Ellis H., 1925-
Drilling engineering handbook.

Includes index.
1. Oil well drilling—Handbooks, manuals, etc.
I. Title
TN871.2.A78 1983 622'.338 82-83470

ISBN 978-90-277-1883-9

Published by D. Reidel Publishing Company,
P.O. Box 17, 3300 AA Dordrecht, Holland
in co-publication with IHRDC

Sold and distributed in North America by IHRDC

In all other countries, sold and distributed
by Kluwer Academic Publishers Group,
P.O. Box 322, 3300 AH Dordrecht, Holland

Contents

Preface

This book presents the fundamental principles of drilling engineering, with the primary objective of making a good well using data that can be properly evaluated through geology, reservoir engineering, and management. It is written to assist the geologist, drilling engineer, reservoir engineer, and manager in performing their assignments. The topics are introduced at a level that should give a good basic understanding of the subject and encourage further investigation of specialized interests.

Many organizations have separate departments, each performing certain functions that can be done by several methods. The reentering of old areas, as the industry is doing today, particularly emphasizes the necessity of good holes, logs, casing design, and cement job. Proper planning and coordination can eliminate many mistakes, and I hope the topics discussed in this book will play a small part in the drilling of better wells.

This book was developed using notes, comments, and ideas from a course I teach called "Drilling Engineering with Offshore Considerations." Some "rules of thumb" equations are used throughout, which have proven to be helpful when applied in the proper perspective. The topics are presented in the proper order for carrying through the drilling of a well.

The help, patience, and understanding of friends, co-workers, and professional associates is greatly appreciated. My special thanks to Field Roebuck, Roebuck-Walton, Inc., Dallas, Texas, for his suggestions, proprietary material, and encouragement. A debt of gratitude to my wife, Virginia, without whose help this book would not have been written.

Drilling Engineering Handbook

1

Principles of Oilwell Drilling

1.1 Introduction

Every oilwell drilling rig must be equipped with systems that enable the rig to meet seven separate but interrelated requirements:

1. Penetrate the subsurface strata.
2. Excavate the drill cuttings.
3. Prevent the caving of penetrated strata.
4. Penetrate deep enough to reach the target reservoir.
5. Drill a hole large enough for efficient production of the reservoir fluids.
6. Keep the hole oriented in the desired direction.
7. Prevent the intrusive fluids from entering the hole.

In modern drilling rigs this is achieved by means of four separate systems:

1. a power plant and transmission system;
2. a hoisting system;
3. a rotating system; and
4. a circulating system.

We shall give a brief description of each of these systems as well as of the personnel, equipment, and materials involved with these systems.

1.2 Power Plant and Transmission System

On most modern drilling rigs the power is supplied by gas or diesel engines or by a diesel-electric system. In direct-drive gas or diesel engines, the power is transmitted to other systems through clutches to a compound which, in turn, is connected by chain drives or by fluid couplings with torque converters. Pumps are usually driven by belt drives. Figure 1.1 illustrates a typical multiengine chain-drive power transmission system.

Diesel-electric rigs use diesel engines to generate electrical power which, in turn, drives individual electric motors on the hoisting, rotating, and circulation systems. Such a system is illustrated in figure 1.2.

Shallow- or moderate-depth rigs (to 5,000 ft or so) require from 500 to 1,000 hp, whereas, heavy-duty rigs require up to 3,000 hp or more. An additional power plant of 100 to 500 hp often provides auxiliary power for lighting, mixing equipment, logging units, etc.

We must mention the SCR rig. SCR means *Silicone Control Rectifier*. Its function is to convert AC voltage to DC voltage by allowing AC voltage to phase through a SCR hockey puck that opens and closes at a very high speed acting like a static flapper valve.

Why not use straight AC current? The reason is that with DC voltage the speed of the electric motor can be varied from 0 to 125 rpm or to 2200 rpm, depending on the demand. Thus, by using SCR power the constant-speed DC power can be converted into useful variable-speed DC power. With SCR, the load efficiency is about 87% as compared to 75% for mechanical rig and 88% for DC–DC rig.

The DC–DC rig requires each motor to have a generator to

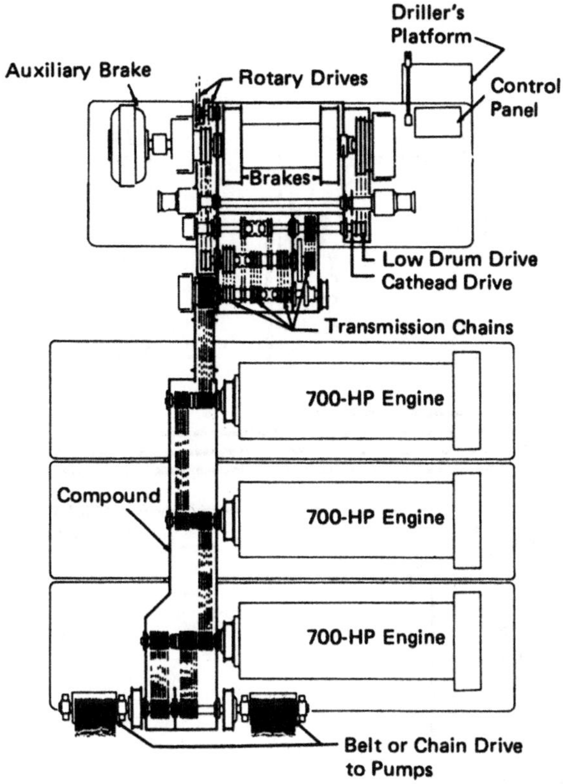

Figure 1.1 Multiengine chain-drive power transmission system. (Courtesy of PETEX, The University of Texas at Austin)

power it and it still requires an AC generator for the rig AC power. The AC load is higher since DC motors and generators require forced-air cooling.

1.3 Hoisting Equipment

The hoisting system is used primarily to raise and lower the drill-pipe in the hole and to maintain the desired weight on the bottom.

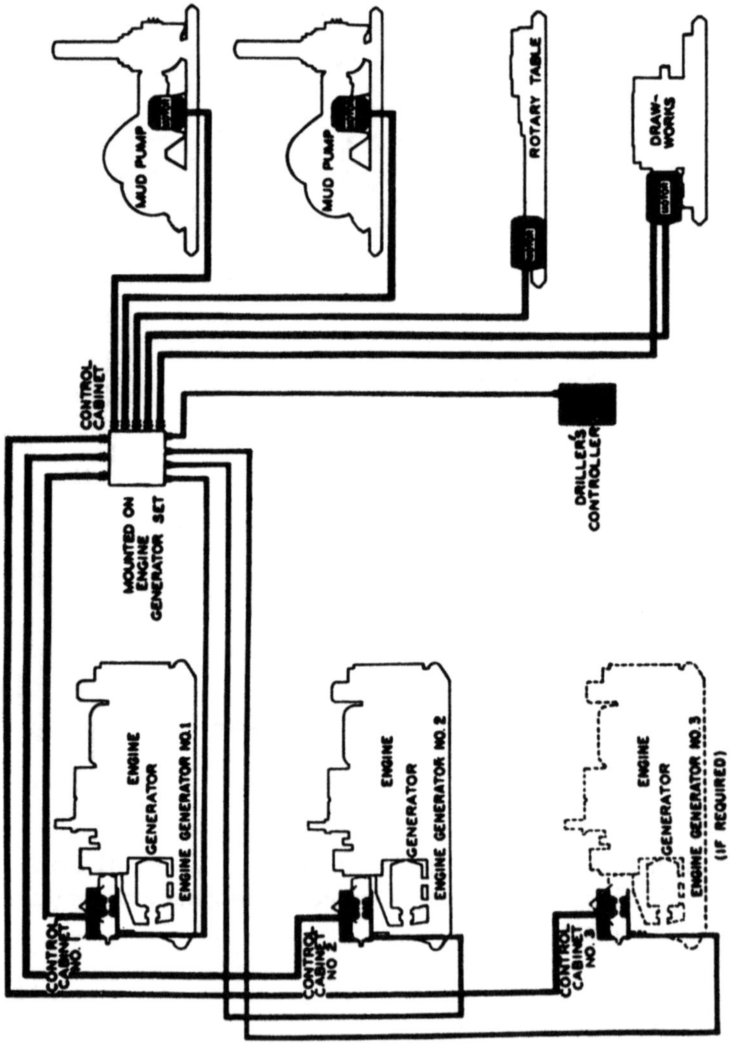

Figure 1.2 Diesel-electric system for power and transmission. (Courtesy of PETEX, The University of Texas at Austin)

As illustrated in figure 1.3, its chief components, other than the derrick or mast itself, include the drawworks, the crown block, and the traveling block-and-hook assembly.

The drawworks is a powered storage drum or reel, around which the fast end of the drilling line is wound. From there, the drilling line passes through the crown and traveling blocks to form a classic block-and-tackle assembly. The dead end of the drilling line then passes through a fixed anchor to the line storage reel.

This system is designed to handle various weights up to some 500,000 pounds or more of dead weight on the hook, depending upon the depth capabilities of the rig, Derrick load capacities vary from 250,000 to 1,500,000 pounds and can stand wind loading from 100 to 130 miles per hours.

The drawworks includes a series of shaft-mounted clutches and chain and gear drives for speed changes and reverse, as well as a main brake with the capacity to stop and sustain the weights imposed during drilling operations.

On each end of the drawworks, a so-called cathead is mounted. One of the types consists of a flanged, spinning drum around which a soft rope is wound to perform several operations. These include light hoisting and assisting in making up and breaking out joints of pipe. The other type of cathead is essentially a manual or air-activated, quick-release friction clutch to which the spinning chain or tong jerkline is attached. This is used exclusively for spinning up and tightening or breaking out drillpipe joints while making trips or adding joints.

Most drilling rigs also include one or more auxiliary power hoists that can be used for light hoisting with wire lines strung over sheaves in the crown block or on a derrick frame member.

1.4 Rotating Equipment

The rotating equipment is used solely to provide a rotating motion to the drill bit at the bottom of the hole. The system includes the rotary swivel, the kelly, the rotary table and kelly bushing, the drillpipe and drillcollars, and the drill bit, as shown in figure 1.4.

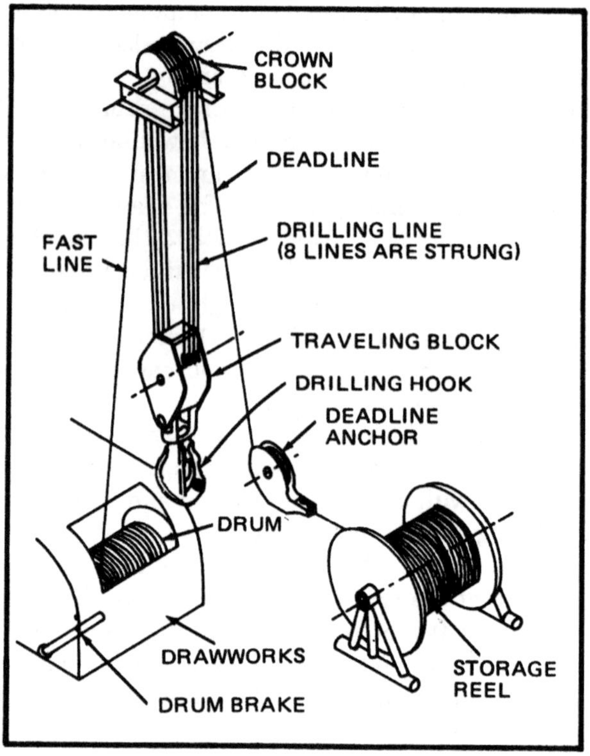

Figure 1.3 Rotary rig hoisting system. (Courtesy of PETEX, The University of Texas at Austin)

Power is transmitted by a chain or gear-driven rotary drive to the rotary table on the rig floor. The square-shaped kelly bushing fits into a hole in the center of the table and turns with it, thereby applying a torque to a square or hexagonal kelly joint which is free to slide vertically through the bushing. The kelly, generally about 40 ft long, constitutes the uppermost joint of the drill string. On top of the kelly is a swivel, which attaches to the hook on the traveling block by means of bails, and a kelly cock, a valve which can be closed to prevent backflow through the drillstring.

The drillpipe is attached to the base of the kelly. Below that are one or more so-called drillcollars and, finally, the drill bit.

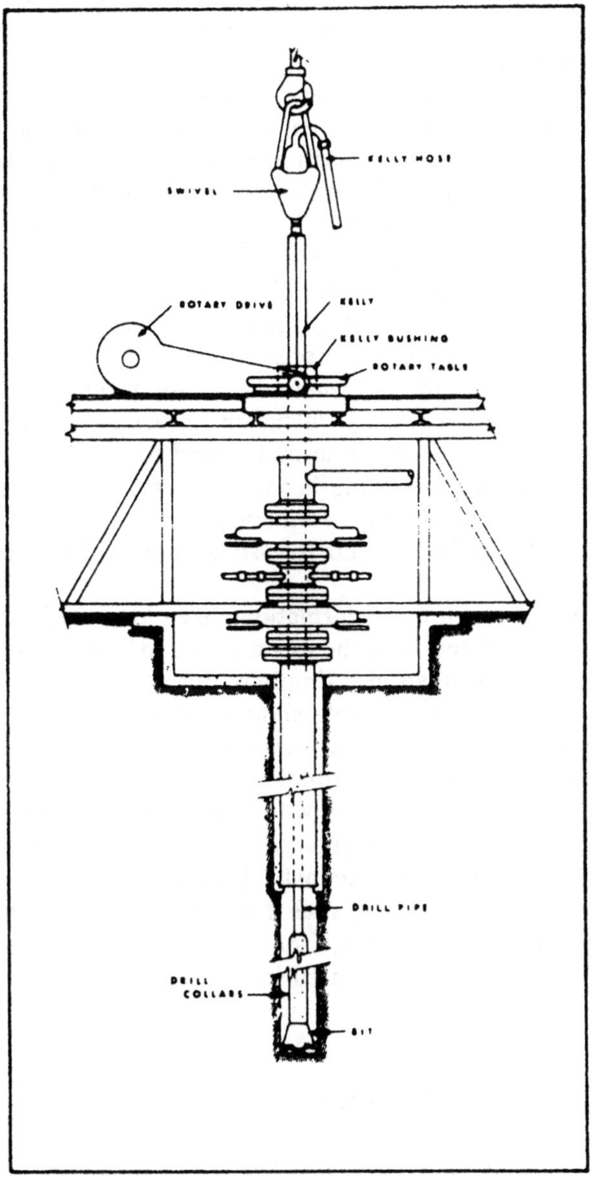

Figure 1.4 Rotary system. (Courtesy of PETEX, The University of Texas at Austin)

Drillcollars are heavy, sometimes especially designed, joints of drillpipe which are used to add bit weight and/or assist in drilling a straight hole.

The rotary table contains a lock to prevent the table from turning when desired, such as when pipe connection is being made or broken. The kelly bushing rides up out of the rotary table on the bottom of the kelly so that the rotary bushing can be used as a seat for tapered pipe slips when making or breaking connections of drillpipe or running casing.

Rotary drill bits are essentially all of a similar design; that is, they have three freely rotating cones which bear on the rock and do the actual drilling. The cones are of two basic types, with either milled teeth or with tungsten carbide inserts, which break and remove the rock by intrusion, pressure breaking, and dragging. Toothed bits are normally used to drill soft to moderately hard formations, while insert bits are used on moderately hard to very hard abrasive formations, with considerable overlapping and a current trend toward a more widespread use of insert bits.

Toothed bits for soft formations have long teeth and cones that are skewed and with the angle of the axis offset to create a scraping action instead of a simple rolling action. Bits for harder formations have progressively shorter teeth and larger bearings and are less offset. Insert bits penetrate the formation with more of a crushing action than toothed bits. Again, the exposed length of the inserts varies inversely with formation hardness.

For extremely hard formations and for some coring operations, coneless bits with industrial diamond inserts are used. Although quite expensive, such bits are often capable of considerably more drilled footage than insert-type core bits. Various types of drill bits are illustrated in figure 1.5.

1.5 Circulating System

The drilling fluid, or mud, makes a circuit through the circulating system of the drilling rig, as illustrated in figure 1.6. Mud is mixed at the mixing hopper from the base fluid, usually water, and the bulk materials in the mud house. From there, it goes to the

Typical Three-Cone Toothed Roller Bits

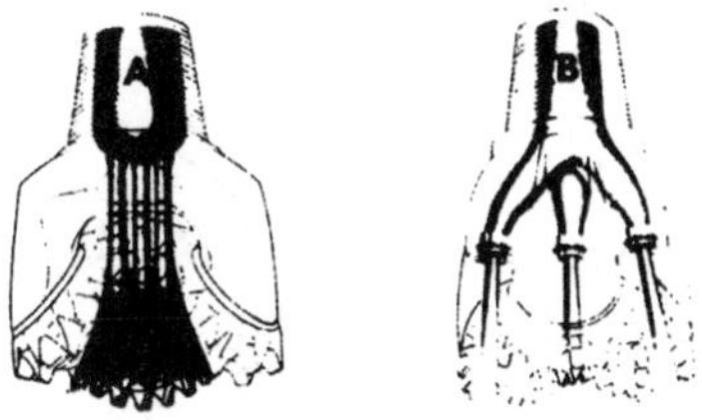

Typical Roller Bit Water Courses

(A) Conventional; (B) Jet Nozzles

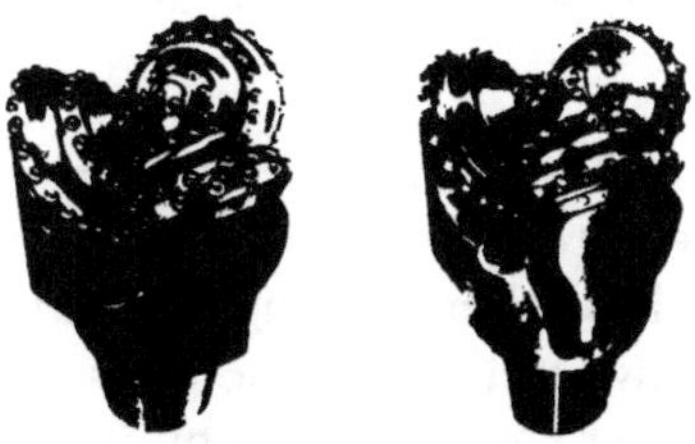

Typical Tungsten - Carbide Roller Bits

Typical Diamond Rotary Bit

Figure 1.5 Types of drill bits.

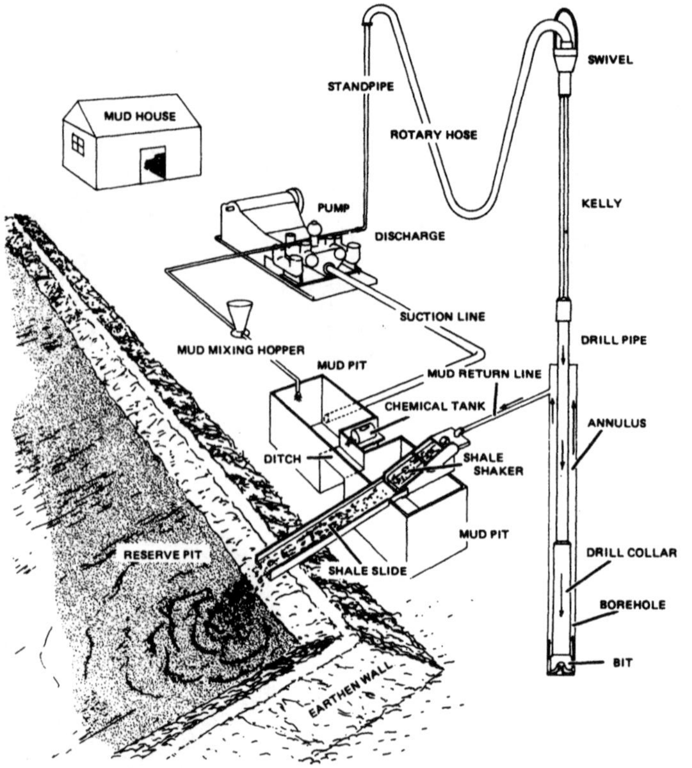

Figure 1.6 Rotary rig fluid circulation and mud treating system. (Courtesy of PETEX, The University of Texas at Austin)

suction pit, where it is picked up by the mud pumps to begin its circuit. It travels up the stand pipe, through the kelly hose (rotary hose or mud hose), and enters the drill column through the swivel. In the drill column, it travels downward through the kelly cock, kelly, drillpipe, drill collars, and water courses of the bit into the drilled hole. There, it picks up drill cuttings and travels with them up the drilled annulus through the blowout preventers and mud-return line to the shale shaker. There, the larger cuttings are screened out, and the mud flows into a settling pit from where it returns through the sump pit to the suction pit to begin the circuit once again.

From the shale shaker, the cutting and contaminated mud are diverted to a so-called reserve pit, which is actually a waste pit rather than a true reserve supply. On most modern rigs, the suction, settling, and sump pits are actually steel tanks rather than the dirt pits of a few years ago.

There are two types of auxiliary systems associated with the basic circulating system, the equipment necessary for well pressure control. The former includes mud pit agitators for maintenance of a uniform content of mud solids, cone-type desanders and desilters to remove contaminants that would not settle out otherwise, and a vacuum degaser for removal of entrained gases.

The equipment for well pressure control includes the blowout preventers under the rig floor on the casing head. These consists of one preventer, which can seal the annulus around the kelly or the drillpipe, and one or more ram-type preventers. Also included here are a connection to fill the hole with mud when making a trip, a kill line, which permits mud to be pumped down the annulus to restore the pressure balance, and annular pressure relief lines. Additional components of this system include the kelly cock and the mud mixer and pumps themselves.

The shale shaker is a system of rotating or vibrating screens, which removes the larger drill cuttings from the returning mud and therefore serves as the sampling point for drill cuttings analysis. Most often the returning mud is also monitored by a logging unit for density and hydrocarbon content. A gauge on the standpipe on the derrick floor permits manual and/or automatic monitoring of mud pumping pressure.

The mud pumps serve as the heart of the system. They are powered by individual electric motors, by belt or chain drives off the main compound, or by belt drive off separate gas or diesel engines. Generally, at least one extra, auxiliary pump is provided for rate and pressure control in the case of mechanical malfunction during critical periods. For the same reason, separate auxiliary power is also often provided.

The information in tables 1.1 to 1.17 is included to help you select equipment and establish operating parameters for drilling wells.

TABLE 1.1 HUGHES PRACTICAL HYDRAULICS RECOMMENDED HYDRAULICS PROGRAM

CONTRACTOR ______ RIG NO. ______ LOCATION ______ DATE ______

OPERATOR ______ FIELD, COUNTY & STATE ______

HOLE SIZE IN. ______ DEPTH INTERVAL FROM ______ FT. TO ______ FT. MAX. SURFACE PRESSURE ______ PSI

DRILL PIPE SIZE ______ WT. ______ TOOL JOINTS SIZE & TYPE ______ O.D. ______

DRILL COLLARS NO. ______ O.D. ______ I.D. ______ NO. ______ O.D. ______ I.D. ______

PUMP NO. ______ MODEL ______ LINER ______ ROD DIA. ______ PUMP VOLUMETRIC EFFICIENCY ______

PUMP NO. ______ MODEL ______ LINER ______ ROD DIA. ______ PUMP VOLUMETRIC EFFICIENCY ______

DEPTH		JET	GPM		PUMP SPM				MIN. ANN. VEL. FPM	MIN. JET VEL. FPS	BIT HYD. HP MIN.	BIT HYD. HP IN.^{2}	IMPACT FORCE MIN.	PUMP INPUT HP MAX.	MUD WT.
					NO. 1		NO. 2								
FROM	TO	SIZE	START	END	START	END	START	END							

REMARKS:

Source: Courtesy of Hughes Tool Division, Hughes Tool Company.

TABLE 1.2 MAXIMUM DESIGN WEIGHT ON BITS (in thousands of pounds per inch of diameter)

	Bit class–subclass								Insert bits				
Bit size	1–1	1–2	1–3	1–4	2–1 2–2	2–3	3	4	5	6	7	8	9
6⅛		5.6	6.0	6.6	6.9		7.9						
6¾		5.7	6.1	6.6	7.1	7.2	8.5		3.1	4.4	4.5	5.2	4.0
7⅞	6.0	6.2	6.6	7.0	7.5	7.6	8.7	9.4	3.5	4.5	5.0	5.7	4.5
8¾	6.2	6.5	6.8	7.2	7.8	8.0	9.5	10.0	3.7	5.1	5.2	5.8	4.7
9⅞	6.5	6.7	7.1	7.0	7.6	7.7	8.9		3.6	5.1	5.1	5.9	4.6
10⅝		6.4		7.0			8.8		3.5	5.0	5.0	5.8	4.5
12¼	5.9	6.1	6.4	6.7	7.3	7.4	8.5		3.5	4.9	4.9	5.6	4.4
14¾		5.3		5.8		6.3	7.4		3.4	4.7	4.8	5.4	4.3
17½		5.0		5.7			7.0		3.0	4.2	4.2	4.8	3.8

1. Sealed bearing bits are 8–10% lower.
2. Journal bearing bits are 10–12% higher.
3. Insert bit maximums are based on cutting structure not bearing capacity.
4. Actual optimum weights are usually 20 to 30% less than these maximums.
5. Rules:
   a) Use the softest-formation bit that will obtain an economical run.
   b) If weight on bit is limited, use the softest-formation bit practical; use the type with the least bottomhole contact.
   c) If the formation is not responsive to weight on bit, use the softest-formation bit practical.
6. Weight in *excess* of that listed above will generally reduce bit life to approximately 6 hours.

TABLE 1.3 BIT AND HYDRAULIC SUMMARY

Gross Interval footage	Hole size	Bit type	Bit depth out	Bit weight		Rotary speed		Velocity annulus	gal/min	Nozzles	Bottomhole assembly[1]
				min	max	min	max				
0–75		Drive									
75–1,000	22″	Hughes OSC–3A Smith DS	1,000	5	20	120	150	45	600 to 850	Conv.	9½″ Drill collars + stab. at 60′
1,000–6,900	14¾″	Hughes OSC–3A Reed Y–11 Smith D–S–J Security S3–J	6,900	30	72	80	150	75 to 105	560 to 880	3–16 2–20–B	Soft shoe or equal and No. 3 assembly
6,900– 11,650	9½″	Hughes J–22 Security S–86–F Smith F–2 Reed FP 52	11,650	40	52	*50	60	90 to 150	260 to 430	3–12 2–15–B	Soft shoe or equal and No. 3 assembly
11,650– 16,000	6½″	Hughes J–33, J–44 Smith F–3, F–4 Security S–86–F M–84–F Reed FP–52, FP–72	16,000	30	35	**50	60	130 to 180	160 to 220	3–9 2–11–B	Soft shoe or equal and No. 3 assembly

*Run drill-off test with each bit change to insure proper weight, best bottomhole hydraulics, and penetration rate.
**All Journal bearing bits will have a maximum rotary of 60 rpm.
[1] See figure 7.2.

TABLE 1.4 WORKSHEET FOR WELL PROGRAM

Depth (m)	ft	Formation top & type	Drilling problems	Type of formation evaluation	Hole size	Casing size depth	Fracture gradient	Formulation pressure gradient	Mud weight type
(309)	1,000								
(617)	2,000								
(926)	3,000								
(1235)	4,000								
(1543)	5,000								
(1852)	6,000								
(2160)	7,000								
(2469)	8,000								
(2778)	9,000								
(3086)	10,000								
(3395)	11,000								
(3704)	12,000								
(4012)	13,000								
(4320)	14,000								
(4630)	15,000								
(4938)	16,000								
(5247)	17,000								
(5555)	18,000								

TABLE 1.5 WELL COST BREAKDOWN (%)

Payments to drilling contractors	36.6
Purchased items	
road & site preparation	4.1
transportation	3.9
fuel	1.1
drilling mud & additives	6.9
well site logging and/or monitoring system	1.2
all other physical tests	0.7
logs & wireline evaluation services	3.2
directional drilling services	0.6
perforate	1.1
formation treating	3.0
cement & cementing services	3.7
casing & tubing	17.5
casing hardware	0.7
special tool rentals	3.1
drill bits & reamers	1.6
wellhead equipment	1.8
other equipment & supplies	2.0
plugging	0.5
supervision & overhead	2.1
all other expenditures	4.6
Subtotal purchased items	63.4%

Source: IPAA Cost Study Committee (1976)

TABLE 1.6 DRILLING VARIABLES

Alterable	Unalterable
a) Mud	1. Location
Type	2. Weather
Solids content	3. Rig conditions
Viscosity	4. Rig flexibility
Fluid loss	5. Corrosive borehole gas
Density	6. Bottomhole temperature
b) Hydraulics	7. Roundtrip time
Pump pressure	8. Rock properties
Jet velocity	9. Characteristic hole problems
Circulation rate	10. Water availability
Annular velocity	11. Formation to be drilled
c) Bit type	12. Crew efficiency
WT on bit	13. Depth
Rotary speed	

TABLE 1.7 DRILLING VARIABLE INTERACTION

Variable Combination	Interaction
WT—RPM	Negative
WT—Hydraulics	Positive
RPM—Hydraulics	None
Low solids—Hydraulics	Positive
Low solids—WT	Positive
Bit type—Formation	Either
Low solids—Bit Type	Positive
RPM—Formation	Negative
Mud solids—Dual-action polymer	Positive

Note that out of the five positive interactions, four are directly related to drilling fluids.

TABLE 1.8 RIG SELECTION FACTORS

1. Depth and hole size.
2. Drilling conditions expected:
   a) Hard rough drilling requires good hoisting capacity and strong drillpipe.
   b) Soft fast drilling requires higher rotary and hydraulic horsepower.
   c) Rig must be equipped to handle expected drilling problems: kicks, lost circulation, deviation.
3. Rig crews efficiency is an important consideration.
4. Condition of rig is very important.

TABLE 1.9 WELL DATA

Depth:	9,000 ft
Size:	7⅞" hole
Formations:	Soft to hard
	300 ft. of rough anhydrite at 2,000 ft
Problems:	Some lost circulation 1,500–4,000′
	20,000–70,000 ppm chlorides to 5,000′
Mud:	9 lb/gal or less to 9,000 ft

TABLE 1.10 RIG RATING CHECKLIST

Rig owner:

Rig identification:

Intended area of use:

1. Capacity for handling pipe ______________________ lb
2. Substructure load supporting capacity
   a) Maximum pipe setback capacity ______________________ lb
   b) Maximum rotary table supporting capacity ______________________ lb
   c) Corner loading capacity (for derricks only) ______________________ lb
3. Hoisting and breaking capacity

   Hook hp Hook load Hook velocity

   Observed hoisting performance: __________ Mlb __________ ft/min

   Auxiliary brake performance: Max hook load ______ Mlb ______ ft/min
4. Mud pump

	High-volume service			High-pressure service		
	Spm	psi	HHP	Spm	psi	HHP
a) Main pump						
b) Stand-by						
c) Stand-by						

5. Rotary table
   a) Make: __________ Model: __________ API opening: __________
   b) Independent drive: max __________ rpm to min __________ rpm
   c) Independent drive unit make: __________ Model __________
   d) Drive unit (engine/motor) make: __________ Model __________
   e) Continuous hp rating: ______________________
   f) Table driven through drawworks

   Ratio I: __________________ rpm to __________________ rpm

   Ratio II: __________________ rpm to __________________ rpm

   Ratio III: __________________ rpm to __________________ rpm

6. Drillstring

	String #1	String #2	
Nominal size	________	________	in
Weight per foot	________	________	lb
API Grade	________	________	
Length of string	________	________	ft
Tool joint size & style	________	________	
Tool joint OD	________	________	in
Date of last inspection	________	________	
Inspection method	________	________	
No. of joints by inspection	________	________	Class 1
	________	________	Premium
	________	________	Class 2
	________	________	Class 3

Other information (such as protectors, hard banding, etc.)

7. Drillcollars

Quantity ______ ______ ______ ______

Max. OD to nearest 1/16 ______ ______ ______

Min. back to nearest 1/16 ______ ______ ______

Average length ______ ______ ______ ______

Approx. weight ______ ______ ______ ______

Tool joint style and type ______ ______ ______

8. Auxiliary equipment

a) Mud tanks:

Number	Size	Capacity (each)
________	________	________
________	________	________

Mud-mixing equipment; Pumps:

Number	Make	Type	Size
________	________	________	________
________	________	________	________

TABLE 1.10 *(continued)*

Prime Mover:

Number	Make	Model	hp
________	________	________	________
________	________	________	________

Mud-agitating equipment: ____________________

Shale-shaker: Make ____________ Model ____________

Desander:

Make	Model	Capacity (gal/min)
________	________	________
________	________	________

Desander pump: Make ______ Type ______ Capacity ______ (gal/min)

Pump prime mover:

Make	Model	hp
________	________	________

9. Blowout prevention equipment

No.	Make	Model	Flange size	Bore	Working pressure
____	______	______	________	______	________
____	______	______	________	______	________

Choke manifold description: ____________________

Kelly cock: Make ____________ Model ____________

Drill pipe safety valve: Make ____________ Model ____________

Degasser: Make ____________ Model ____________

Mud–gas separator descriptions: ____________________

Blowout preventor closing unit: Make ______ Model ____________

No. of stations ____________ No. of outlets: ____________

Accumulator volume: (liquid and gas) ____________________ gal

Precharge pressure (before adding liquid) ____________________ psi

Final pressure (when fully charged with liquid) ____________________ psi

Reducing and regulating valve: Yes ____________ No ____________

10. Generators

Number	Name	Model	Type	Capacity (kw)
________	________	________	________	________
________	________	________	________	________

Generator prime mover

Number	Make	Model	hp
________	________	________	________
________	________	________	________

Lighting system

Vapor proof: Yes ________________ No ________________

Mud storage description: ________________________________

____

____

Cement storage capacity: ________________________________

Freshwater storage capacity: ________________________________

Fuel storage capacity: ________________________________

Additional equipment: ________________________________

____

____

11. Mast or derrick

a) Rig hook load capacity for drill string is ________________ lb
as limited by ________________________ below.

b) Rig hook load capacity for running casing is ________________ lb
as limited by ________________________ below.

c) Wind resistance with no drill pipe ________________ mph.

d) Wind resistance with ________ ft of pipe ________ mph.

e) Racking capacity of ________ inch drill pipe ________ ft.

f) Static hook-load capacity with ________ lines strung ________ Mlb.

g) Crown block: Make ________ Weight ________ No sheaves ________
Tons ________________________

h) Traveling block: Make ________ Weight ________ No. sheaves ________
Tons ________________________

i) Hook: ________________________ Tons

j) Swivel: ________________________ Tons

TABLE 1.10 *(continued)*

k) Elevator links: Size ________ Length ________ Tons ________

l) Drillpipe elevators: ________________________________ Tons

m) Casing elevators: ________________________________ Tons

n) Drilling line (hook load capacity): ______________________ Mlbs

No. of links: ______________ Classification: ______________

Size: ___________ Grade: ___________ Center: ___________

API breaking strength: ___________ Safety factor: ___________

12. Substructure

Make: ____

Model or type: ________________________________

Floor height: ________________________________ ft

Width: ______________________________________ ft

Length: _____________________________________ ft

Height above ground to underside of rotary beam _______ ft _______ in

Maximum pipe setback capacity: ______________________ Mlb

Maximum rotary-table supporting capacity: ______________ Mlb

Corner loading capacity (for derricks only): ______________ Mlb

Loads imposed by tensioning devices: ____________________ Mlb

Additional loads: ___________________________________ Mlb

13. Hoisting

a) Maximum hook-horsepower observed to be ____ at hook load of ____ lb

b) Drawworks: Type ________ Make _______ Model _______ Size _______

c) Auxiliary brake: Type ________________________________

Make ___________ Model ___________ Size ___________

d) Engine-hoist transmission system description: ______________

e) Hoisting engines: ________________________________

f) Observed test data:

No. of link strung	Drawworks ratio	Hook-load (Mlb)	Sec. to pull middle single
________	________	________	________

Hook velocity ft/min.	Cal. hook hp
______	______

g) Observed auxiliary brake performance

Hook load (Mlb)	Seconds to lower middle single	Hook velocity (ft/min)	No. of Lines to traveling block	Remarks
______	______	______	______	______
______	______	______	______	______

14. Mud pump

	Main	Stand-by	Stand-by
a) Make:	______	______	______
b) Model:	______	______	______
c) Rated hp:	______	______	______
d) Rated rpm and stroke:	___×___in.	___×___ in.	___×___ in.
e) Liner size max − min:	______	______	______
f) Working press:	______	______	______
g) Engine make:	______	______	______
h) Engine model:	______	______	______
i) hp at_____ rpm_____	______	______	______

j) Do same engines drive rotary? ______________________________

____

TABLE 1.11 DRILLING RIG DATA

Equipment	Rig A	Rig B	Rig C
Drawworks	Oil well 76	National 50	Gardner–Denver 750
Power	2–325 hp 1–175 hp on independent rotary	2–325 hp	2–450 hp
Rotary speeds	High-gear 135 max Low-gear 75 max	120–80–45	300–135–100–75–40
Mud pumps	550 hp main 370 hp stand-by	600 hp main 325 hp stand-by	700 hp main 370 hp stand-by
Drill collars	29–(6¼ × 2¼) (67,000 lb max)	22–(6¼ × 2¼) (51,000 lb max)	27–(6¼ × 2¼) (63,000 lb max)
Other	Shale shaker Chemical mixer	Shale shaker Chemical mixer Premix mud tank	Shale shaker Chemical mixer Premix mud tank

TABLE 1.12 RIG SELECTION

Specification	Rig A	Rig B	Rig C
Bid day rate	6700	6400	6100
footage	—	19.50	18.50
Sufficient rotary	Not enough	No	Yes
Power and selection	Power with weight		
Adequate pump and power 550 hp min	Marginal	No Not enough power	Yes
Adequate drill-collars max of 63,000 lbs needed	Yes	No	Marginal
Adequate mud	No	Yes	Yes
Handling equipment	Premix tank is needed for high chloride content		

TABLE 1.13 ANALYSIS OF OPERATION

Rig A	Insufficient rotary horsepower and marginal pump to handle drillcollars
Rig B	Not enough drillcollar weight, insufficient horsepower to handle more. Plant to use clear water for drilling fluid to 5200 ft.
Rig C	Balanced rig—needs more drill collars in certain sections of hole.
Actual case:	
Rig A	Took 24 days to drill hole
Rig B	Took 21.5 days to drill hole
Rig C	Took 19 days to drill hole

TABLE 1.14 DRILLING EQUIPMENT DESIGN

1. Bursting pressure of cylinder:

$$BP = \frac{2TS}{D},$$

where

BP = lbs/in. pressure,

S = tensile strength of material,

T = wall thickness,

D = outside diameter.

2. Discharge pressure:

$$WP = \frac{\text{hp} \times 1455}{\text{Volume gpm}},$$

$$\text{hp} = \frac{\text{psi} \times \text{Volume gpm}}{1455},$$

$$\text{Volume gpm} = \frac{\text{hp} \times 1455}{\text{wp}}$$

where

wp = discharge pressure,

hp = horsepower,

gpm = gallons per minute,

psi = pounds per square inch.

TABLE 1.14 *(continued)*

3. Lubrication for chain: 1/2 pint per strand per minute for 1 inch

EXAMPLE Three six-strand drive 1¾ pitch,
18 strands = 9 pints,
1¾ pitch = 15 pints = 1.9 gal/min.

4. Chain Speed:

$$\mathrm{CS} = \frac{T \times \mathrm{rpm}}{K},$$

where T = number of teeth of driving sprocket,
K = ⅜″—32, ½″—24, ⅝″—19.2, ¾″—16, 1″—12, 1¼″—9.6, 1½″—8, 1¾″—6.85, 2″—6, 2½″—4.8,
rpm = revolutions per minute.

Recommended chain speed of 3000 ft/min.

5. Working load of chain:

$$\mathrm{WL} = \frac{\mathrm{hp} \times 33{,}000}{\mathrm{ft/min}} \quad \text{or} \quad \mathrm{WL} = \frac{\mathrm{hp} \times 396{,}000}{T \times P \times \mathrm{rpm}},$$

where T = number of teeth,
P = chain pitch.

6. Chain pull:

$$\mathrm{CP} = \frac{126{,}000 \times \mathrm{hp}}{\mathrm{rpm} \times D},$$

where D = pitch diameter of driving sprocket.

7. Permissible chain pull:

$$\mathrm{PCP} = \frac{5{,}000{,}000 \times A}{V + 575},$$

where A = projected rivet bearing area,
V = chain speed.

PROJECTED RIVET BEARING AREA

Pitch	*A*	Pitch	*A*
3/8	0.040	1½	0.603
½	0.068	1½	0.632
5/8	0.107	1¾	0.722
¾	0.161	1¾	0.751
¾	0.175	2	0.987
1	0.275	2	1.022
1	0.294	2½	1.667
1¼	0.399	2½	1.978
1¼	0.423		

8. Allowable hp per strand of chain:

$$\text{AHP} = \frac{\text{Chain pull} \times \text{Velocity}}{33{,}000}.$$

9. Design speed of drawworks:

   1 stand—90 ft/min

   Fast-line speed—3000 to 3500 ft/min

   Design second speed on four speed (electric)

   Design third speed on six speed (mechanical)

10. Line speed on rigs (ft/min):

$$\text{LS} = C \times \text{rpm}_d$$

where C = Circumference of drum at second layer of line (ft),

rpm_d = revolutions per minute of drum.

11. Line pull on rigs (assume 85% efficiency):

$$\text{LP} = \frac{63{,}025 \times \text{hp} \times N}{\text{rpm}_d \times \text{RD}}$$

where LP = line pull,

N = number of engines,

hp = hp of each engine,

rpm_d = revolutions per minute of drum.

12. Hook pull:

$$\text{HP} = \text{line pull} \times \text{number of lines}.$$

TABLE 1.14 *(continued)*

13. Ton miles on wire line (per round trip)

$$TM = \frac{WL(I+L) + 4IM + 12}{5{,}280 \times 2000},$$

where L = length of pipe string,

I = length of each stand,

W = weight of pipe (lb/ft),

M = weight of blocks, hook, etc. (10,000#).

14. Torsional stress on plain shaft:

$$S = \frac{M}{22},$$

where M = torque (in.-lb),

s = section modules on shaft.

15. Torsional stress on keyway shaft:

$$S = \frac{16M}{\pi B^2 H},$$

where M = torque (in.-lb),

π = 3.1416,

B = keyway width,

H = keyway depth.

16. To determine engine horsepower:

$$HP = \frac{P \times A \times L \times 2 \times R \times \text{no. of cylinders}}{33{,}000}$$

where P = mean effective pressure on piston (use intake manifold pressure),

A = area of piston (sq. in.) less ½ cross-sectional area of piston rod,

L = length of stroke in feet,

2 = number of strokes per revolution of one piston,

R = rpm of engine.

17. Horsepower of engine:

$$HP = \frac{\text{BMEP} \times \text{Disp} \times \text{rpm}}{792{,}000},$$

where BMEP = brake mean effective pressure,

Disp = displacement of engine (cu. in.),

$$\text{HP @ 80 BMEP} = \frac{\text{Disp.} \times \text{rpm}}{10{,}000}.$$

18. Capacity of dry-disc friction clutch:

$$\text{HP} = \frac{F \times \text{MR} \times A \times N \times P \times S}{63{,}000},$$

where F = coefficient of friction (approx. 40),

MR = mean radius of friction line pull,

A = area of friction line pull (sq. in.),

N = number of surfaces,

P = pressure,

S = rpm of shaft.

19. Torque required at rotary clutch to cause pure torsional failure of pipe:

$$T = 26{,}700 \frac{1}{\text{TR}} \times \frac{\text{LSS}}{\text{RTS}} \times \frac{\text{CS}}{\text{LsCs}},$$

where TR = table ratio,

LSS = lineshaft sprocket to rotary,

RTS = rotary table sprocket,

CS = clutch sprocket,

LsCs = lineshaft to clutch sprocket.

20. Horsepower developed by brake to stop load of pipe:

$$\text{HP} = \frac{W \times L \times 60}{33{,}000 \times T},$$

where W = weight of pipe total string (lb),

L = length of pipe (lowered),

T = fall of pipe (ft/second).

TABLE 1.15 DRILLING RULES OF THUMB

1. Mud flow rate:	30 to 50 gpm/in. of bit diameter,
2. Jet horsepower:	2½ to 5 hp/in.^{2} of bottom area,
3. Maximum bit hydraulic:	50 to 65% of available pump pressure across the jet nozzles,
4. Number of jets:	two jets for smaller bit at low penetration rates; otherwise three,
5. Optimum penetration:	4 to 4½ hp/in.^{2} of bit size with nozzle velocity 350 to 400 ft/second.

TABLE 1.16 RIG DESIGN RULES OF THUMB

1.	Mechanical drawworks (with overall efficiency of 75%) generally has a rated load of 267 lb/hp. Thus a 1000 hp rig would be rated at 267,000 lbs.
2.	Drilling rig power: 10 ft/hp
3.	Engine weight: 16 lb/hp
4.	Motor weight: 7 lb/hp

TABLE 1.17 DISTRIBUTION OF RIG POWER WITH MAXIMUM DRAWWORKS 2000 hp

	Drilling condition	Stuck pipe	Running casing	All maximum
Drawworks	0	2000	2000	2000
Rotary	800	0	0	800
Slush pumps	2600	1300	0	2600
Accessories	400	200	200	400
Housekeeping	100	100	100	100
Total:	3900	3600	2300	5900

1.6 Rig Personnel

The typical drilling rig crew consists of a driller, two or three floormen or "roughnecks", a derrickman, and perhaps a motorman. The drilling operation is supervised by a drilling supervisor, more often known as a drilling foreman or tool pusher, augmented by a drilling engineer, a wellsite geologist, a mud engineer and a mud logger.

The floormen do the hard work on the rig floor; the work consists mainly of making and breaking connections and adding or removing joints of drillpipe, while the derrick man assists the operation on the "monkey board" 60, 90, or 120 feet up in the derrick. The driller controls all the rig functions and handles the brakes from his console at one side of the drawworks. Typically, the line of advancement is from floorman to derrickman to driller or tool pusher.

A single drilling crew normally works an 8-hour tour (pronounced "tower") each day. However, in remote or offshore locations, a crew may work a 12-hour tour. When running casing, a special casing crew is called in, or a drilling crew will work overtime to assist the next tour's crew in the operation.

Typically, the tool pusher and the drilling crew will be employees of a drilling contractor, whereas the drilling engineer and wellsite geologist will be employees of the producing company or consultants. At times, the tool pusher or drilling foreman will be a producing company employee. Mud logging personnel and the mud engineer are normally service company employees.

2
Drilling Fluids

2.1 Functions

Although the drilling of a well is a complex operation involving many different mechanical elements and processes, the single most important factor upon which the successful completion of the well depends is the drilling-fluid circulation system. The majority of serious problems encountered during drilling, including lost circulation, stuck pipe, kicking wells, poor penetration performance, high costs, blowouts, and poor-quality well logs, can all be traced back to poorly designed, misunderstood, and misused drilling-fluid systems.

The primary functions of the drilling fluid and its circulation systems are:

1. To remove rock cuttings from the bottom of the hole so that the bit can drill on a fresh rock surface, thereby increasing the efficiency of the drilling operation.
2. To transport the cuttings to the surface where they can be removed from the drilling fluid.
3. To suspend the cuttings in the hole whenever mud circulation is stopped.

4. To cool and lubricate the bit and clean its cutting surface.
5. To exert sufficient hydrostatic pressure to exclude formation fluids from the hole.
6. To maintain a stable, lubricated wellbore that can be reentered at any time during the drilling operation.

Of secondary importance, but critical to reservoir evaluation and control, is the requirement that the drilling should be conducive to obtaining reliable logs of good quality and should have filtration properties to prevent contamination of productive strata. By the same token, it should not deposit a mud cake so thick as to reduce the hole diameter to the point of creating a swabbing action as the drillpipe is reciprocated or causing the pipe to stick.

2.2 Properties of Drilling Fluids

The effectiveness with which a drilling fluid performs its necessary functions depends on its composition and on various physical, chemical and electrical properties. Generally, a drilling fluid can be classified according to its principal fluid phase, i.e., water, oil, or gas. Frequently, the terms *water base* and *oil base* are used to distinguish drilling fluids that have a liquid as the principal component.

A drilling fluid, then, is composed of a fluid phase and of various materials that are added to impart or control the necessary properties, plus contaminating solids and fluids picked up in the hole. The measurable variable properties of a drilling fluid are its density, viscosity, water loss, gel strength, pH, resistivity, and abrasiveness. These properties depend on the size, shape, number, hardness, and composition of the solid materials and on the chemical and interfacial characteristics within the suspension in the wellbore.

TABLE 2.1 DENSITIES OF COMMON MUD COMPONENTS

Material	gram/cm^3	lb/gal	lb/ft^3	lb/bbl	kg/m^3
Water	1.0	8.33	62.4	350	1000
Oil	0.8	6.66	50	280	800
Barite	4.3	35.8	268	1500	4300
Clay	2.5	20.8	156	874	2500
Salt	2.2	18.3	137	770	2200

Reference to the density of water in metric and oil-field units shows that a convenient volume for laboratory or field-pilot tests is 350 cm^3, since grams/350 cm^3 is equivalent to lb/bbl (table 2.1).

2.3 Composition and Treatment of Water Muds

To emphasize what has been said before we shall repeat that the drilling fluid is one of the factors on which the satisfactory completion of the job depends. Several chores are assigned to it, but the relative importance of these assignments depends on local conditions. Certain properties are measured as an indication of expected performance. Materials are added to modify properties in anticipation of improved performance. The objective of the mud engineer is not to set ideal specifications for mud properties, but to complete the well at the lowest overall cost.

Many factors affect the choice of the drilling fluid, and their relative importance can be assessed only for a specific well. In the discussion to follow, no attempt will be made to specify the "best mud" or certain properties, but only to outline some mud compositions that are now being used.

2.3.1 Low-Solids System

If *clear water* fulfills the requirements for the hole stability and cuttings recovery, drilling rate may be improved by: (1) addition of a wetting and detergent agent, 1 part to 1,000 parts of water;

(2) addition of flocculant and provision of a large settling area or mechanical means (desander, desilter) for the removal of cuttings.

If water is adequate for the hole stability, but the cuttings are too small for geological records, cuttings size may be increased by: (1) addition of asbestos, 2 to 3 lb/bbl, or (2) addition of a biopolymer, 0.5 to 1.0 lb/bbl; 0.3 lb/bbl chromic chloride, and 0.2 lb/bbl sodium trichlorophenate if the water is not salty.

When problems arise such that control of filtration properties appears desirable, several choices are available.

Addition of filtrate—reducing polymer to the water (fresh or salt) containing asbestos or biopolymer. Polymers may be high-viscosity CMC, other cellulosic polymers, sodium polyacrylate, or modified starch derivatives. The selection is based on the viscosity and filtration requirements and material costs.

Bentonite extender—vinyl acetate, maleic acid polymer 0.05 lb/bbl; soda ash 0.2 to 0.5 lb/bbl; bentonite 10 to 12 lb/bbl. Properties are maintained by consistent additions of a polymer as drilling proceeds. The amount of bentonite to be added is estimated from a methylene blue test on the mud. Mechanical aids to solids removal are essential to economical operations. The system is not applicable in water containing more than 10,000 ppm chloride.

High-viscosity CMC (sodium carboxyl methyl cellulose) and bentonite can be used in drilling hard rock, but excessive dilution is required in mud-making shales. High-viscosity CMC: 0.5 to 1.0 lb/bbl; bentonite 5 to 10 lb/bbl; fresh water.

Diesel oil can be added to any of the compositions listed above. Emulsifiers may be used if needed, although the emulsification is usually satisfactory without them.

Low-solids muds have been defined as drilling fluids in which suspension and filtration properties are supplied mainly by polymers.

2.3.2 Fresh Water—Bentonite

The starting of "spud" mud is frequently prepared by adding bentonite to fresh water in such amount as is needed to stabilize near-surface soil, loose sand, and gravel. Between 10 and 25 lb/bbl is added initially. If mud-making clays and shales are drilled, water alone may be added until the surface hole has been made. Drilling may continue with this native mud, with or without supplemental additions of bentonite, and thinners may be used to facilitate separation of cuttings (as has been described). The properties of such muds are seriously affected by inclusion of cement, salt, or anhydrite. The repulsive forces between the clay particles are reduced by the electrolyte. Base exchange of calcium for sodium occurs with contamination by cement and anhydrite. Edge-to-face and edge-to-edge association of particles causes the mud to "clabber." Apparent viscosity and gel strength rise rapidly, as well as filtration rate.

Treatment of the contaminated mud is determined largely by the requirements for subsequent drilling. The incidence of cement contamination is always known, and the decision can be made as to whether to discard the mud, to treat it with sodium bicarbonate and thinners, or to treat it heavily with chrome lignosulfonate. Salt may enter the mud through the saltwater flow or by drilling rock salt. A saltwater flow will be brought under control before drilling is resumed; accordingly, addition of caustic soda, chrome lignosulfonate, and perhaps polymers to reduce filtration will restore satisfactory properties. Rock salt must be drilled with salt-saturated mud to avoid hole enlargement; consequently, preparation of salt-saturated mud at this time is recommended.

The thickness of the anhydrite section will determine the course of treatment of the anhydrite-contaminated mud. If the section is thin, treatment with soda ash will precipitate the calcium; or barium carbonate will precipitate calcium and sulfate; or the complex phosphates will sequester the calcium ion. If the anhydrite section is thick, a more economical program is to proceed with a calcium-treated mud, i.e., to accept the calcium sulfate as a normal component of the system—no longer to be a "contaminant." Chrome lignosulfonates and caustic soda are added to make the transition easy. Treatment consists of chrome lignosulfonate 4 to 6 lb/bbl; caustic soda 1 to 3 lb/bbl to raise pH to 9.5; and CMC or modified starch polymer to reduce filtration if required.

2.3.3 Chrome Lignosulfonate—Freshwater Mud

If the native shale mud has been discarded after surface casing has been set, a freshwater mud for deeper drilling may consist of the following: bentonite 10 to 15 lb/bbl, chrome lignosulfonate 4 to 6 lb/bbl, and caustic soda 0.5 to 1.0 lb/bbl to raise pH to 9.5 or 10.5. As drilling progresses and the need for better hole-stabilizing properties becomes evident, the chrome lignosulfonate content is raised to 8 to 10 lb/bbl, and lignite or chrome-lignite (4 to 6 lb/bbl) is added. Caustic soda is added as needed to maintain pH at 9.5 to 10.5. As bottomhole temperature rises, lignite additions are increased.

2.3.4 Calcium-Treated Muds

Prior to the introduction of ferrochrome lignosulfonate, *lime muds* were widely used because of their tolerance of the usual contaminants (including shale solids). These systems were characterized as (1) high-lime high-alkalinity and with decreasing undissolved lime and lower alkalinity, (2) low-lime low-alkalinity. The low-lime low-alkalinity muds were more affected by contaminants but were less likely to solidify at high temperatures. Lime muds are rarely used now, but occasionally a severely cement-contaminated mud becomes a lime mud after treatment with chrome lignosulfonate.

2.3.5 Gypsum-Treated Muds

Gypsum-treated muds, which were made practical by ferrochrome lignosulfonate, as stated earlier, offered several significant advantages over lime muds. These advantages included stability at higher temperatures, easier maintenance of desired properties, and significantly higher calcium-ion concentration to repress dispersion of shale. Usually gyp mud was made from a freshwater mud by the addition of chrome lignosulfonate 4 to 6 lb/bbl, gypsum 4 to 8 lb/bbl, caustic soda 0.5 to 1.0 lb/bbl, (pH 9.5 to 10), and CMC 0.5 to 1.0 lb/bbl. Currently gyp muds usually result from, or are prepared in preparation of, drilling anhydrite, as was discussed above.

2.3.6 Saltwater Muds

Saltwater muds can originate in consequence of the make-up water or by addition of salt. If the make-up water contains more than 5,000 ppm chloride, the suspending and sealing properties of bentonite are seriously impaired. Brackish-water muds usually have about the same component as freshwater muds but contain larger amounts of thinners and organic polymers.

In seawater (approximately 15 to 20,000 ppm chloride) the presence of salts of magnesium and calcium inhibit the hydration and swelling of clays in the formations drilled. Magnesium-ion concentration is greatly reduced when pH is above 10; therefore, pH should be controlled between 9.0 and 9.5 when chrome lignosulfonate is used as a deflocculant.

As with freshwater muds, the composition selected for a saltwater mud depends on the specific application. Guar gum alone (0.5 to 1.5 lb/bbl) may provide adequate suspension and filtration properties in workover operations. Asbestos (1 to 3 lb/bbl) or attapulgite clay (15 to 20 lb/bbl) affords suspension of cuttings in water of any salinity but filtration control must be furnished by organic polymers.

When some fresh water is available at the drillsite, prehydrate bentonite slurry can be added to the salt water. The slurry is prepared by thoroughly mixing 30 to 35 lb of bentonite per bbl of fresh water containing 4 lb of chrome lignosulfonate and 0.5 lb of caustic soda. The slurry is added to salt water in the ratio of 1:3. The chrome lignosulfonate not only aids in dispersion of the bentonite but also contributes substantially to maintenance of properties by protecting against the dehydrating action of the dissolved salt. Prehydrated bentonite can be used with water of any salinity. Supplemental treatment with organic polymers provides filtration control as needed. In general, the higher the salt concentration, the more polymer is required.

Frequently, the most economical composition for a salt-saturated mud consists of attapulgite clay, 10 to 15 lb/bbl, and starch, 4 to 8 lb/bbl. When a salt-saturated mud is made from a freshwater mud, pilot tests should be made to estimate the amount of dilution that will be necessary before salt (approximately 125 lb/bbl) is added. Salt saturation is not maintained, fermentation of starch is likely. Paraformaldehyde (0.3 to 0.5 lb/bbl) is an effective preservative. Sodium pentachlorphenate is also effective. Normally saturated salt muds are used in drilling bedded or dome salt. If calcium and magnesium salts are present along with sodium chloride, coprecipitation of starch will occur if soda ash or caustic soda is added to remove the alkaline earth metals. Diammonium phosphate can be used instead of the alkalies.

Saltwater muds usually foam, and air entrapment may become so severe as to hinder the operation of the mud pumps. Several defoamers are in common use; the most effective can be established only by trial in the particular system. The defoaming agent can cause release of air only at the surface of the liquid; consequently, minimum viscosity and gel development favor breaking of foam. Aluminum stearate (dissolved in diesel oil), octyl alcohol, alkyl aryl sulfonates, and sulfated castor oil are the principal commercial defoamers. The amount needed is small; the problem is to apply it in such a way as to expose the maximum surface of mud to its action.

2.3.7 Nondispersed Weighted Mud

The deleterious effect on drilling rate of increased solids in the drilling fluid has been amply documented. As downhole pressures increase, density of the mud must be raised and solids content must increase. Efforts have been directed, therefore, toward holding the solids to a minimum. The approach has been to try to avoid incorporation of drill cuttings into the mud system. Very little, if any, alkalies and dispersants are added; polymers are used to give suspension properties and mechanical aids are employed for solids removal.

In one nondispersed weighted mud, a maleic acid-vinyl acetate copolymer serves the dual purpose of thickening the added bentonite and flocculating drill solids. The amount of bentonite needed for suspension of the barite decreases as the density increases and is held at the minimum that will supply the needed properties. The total solids content of the mud is used as a control, along with the methylene-blue test as a measure of active clay content, and steps are taken as needed to hold the ratio of inactive solids to bentonitic solids at less than 2:1. In preparing the weighted mud, an approximate composition is: soda ash to reduce hardness to about 200 ppm; bentonite, 5 to 15 lb/bbl; maleic-acid vinyl-acetate polymer, 0.05 to 0.1 lb/bbl; and 5 lb of polymer per 10,000 lb of barite added to raise the density. As drilling progresses, the program of treatment is based on the mud-making qualities of the formations drilled, as shown by base exchange capacity and retort tests. Polyacrylate polymer is added if lower filtration rate is wanted.

2.3.8 Potassium-Treated Muds

So-called potassium-based muds contain potassium chloride and may contain other potassium compounds, such as potassium hydroxide and organic-potassium compounds. The principal purpose of potassium muds is to stabilize troublesome shales. Such muds are characterized as nondispersed; they show a ratio of yield point to plastic viscosity from 3:1 to as high as 8:1. Prehydrated bentonite and organic polymers provide suspension and filtration properties. Filtration rate is not of concern in drilling shale.

The mechanism proposed for the stabilization of shales by potassium muds is as follows: when montmorillonite is present in the shale, potassium exchanges for sodium and calcium because of its preferred ionic size and low hydrational energy. The potassium montmorillonite will have a less hydratable structure than the original mineral. The potassium replaces any exchangeable bases in illite. With interlayered minerals, potassium acts on both segments and reduces the amount of differential swelling that can occur.

Several potassium mud compositions have been used:

a) Bentonite, 5 to 15 lb/bbl, prehydrated; potassium chloride 10 lb/bbl; acrylamide polymer with a molecular weight in excess of 3,000,000, 0.25 to 0.5 lb/bbl; caustic soda, 0.2 to 0.3 lb/bbl to adjust pH to 9.5 or 10.5.
b) Bentonite, 0 to 10 lb/bbl, prehydrated; potassium chloride, 15 to 30 lb/bbl; X–C polymer, 0.2 to 0.5 lb/bbl, with or without chromic chloride for cross-linking; caustic potash to adjust pH to 9.5 or 10.
c) Bentonite, 7 to 10 lb/bbl, prehydrated; potassium chloride, 10 to 20 lb/bbl; potassium lignite, 20 to 35 lb/bbl; nonionic surfactant, 1 to 8 lb/bbl; in weighted muds, biopolymer 0.1 to 0.3 lb/bbl, and reduce bentonite accordingly.

Defoaming agents, such as described for salty muds, usually are required with potassium muds.

2.4 Oil-Base Muds

Drilling fluids that contain oil as the continuous liquid phase are called *oil-base* or *oil muds*. Such muds always contain some water, and if the water is emulsified as a useful constituent, the mud is called an *invert-emulsion mud.*

2.4.1 Applications

Principal applications for oil muds are: (1) to prevent damage to the productive formation by the drilling fluid; (2) to drill or core evaporites; (3) to drill troublesome shales; (4) to overcome wall-sticking of drillpipe; (5) to release stuck pipe; (6) to drill under extreme temperature conditions, high temperatures in very deep holes and low temperatures in permafrost and cold climates; (7) to place in the tubing-casing annulus and the casing-hole annulus to facilitate recovery of pipe; and (8) to drill formations containing corrosive fluids, such as hydrogen sulfide.

2.4.2 Components

Oil makes up 60 to 98% of the liquids in oil muds. Diesel fuel is commonly used, although some crude oils are satisfactory. For reasons of safety, the flash point of the oil should be above 160°F. The analine point should be at least 150°F to minimize damage to rubber parts. Oil-resistant rubber should be used wherever rubber parts come in contact with oil mud. Winter-grade diesel oil is required in extremely cold weather.

Water, the dispersed or emulsified phase, is present in amounts of 2 to 40% by volume. Between 15 and 30% is normal for invert-emulsion muds. Water from almost any source is acceptable (an exception is produced water that contains emulsion breakers) because the chemical composition of the water usually is adjusted for the particular application of the oil mud. For example, calcium chloride is added to the water to improve the hole stability in shale.

The other components of oil muds are varied. Often the oil mud is mixed at a central mixing plant and delivered to the well site where barite is added if needed. Although the composition differs among the several commercial oil muds, the constituents serve to provide the properties necessary for: (1) suspension, such as organophilic clays, asphalt; (2) emulsification, such as calcium soaps, may be formed in the system by reaction of quick lime and fatty acids; (3) filtration, such as asphalt, resins, lignite derivative; (4) oil wetting, such as lecithin; (5) shale stabilization, such as calcium chloride, salt; (6) viscosity reduction, such as petroleum sulfonates; and (7) increase density, such as limestone, barite.

Special techniques are required for some of the tests that are routinely made on oil muds. These tests include chloride, alkalinity, and calcium.

Each of the proprietary oil muds has been devised to give satisfactory performance when made from brand-name product. The composition usually is secret. Some ingredients of one oil mud may be compatible with another; others may not. Formulation may consist of a mixture such as:

Emulsion of mud: mix 70% water with 0.5% soap; add 30% diesel to this mixture.

Inverted emulsion of mud: mix 80% diesel with 0.5% soap; add 20% water to this mixture.

2.4.3 Shale Stabilization by Oil Mud

An obvious solution to hole problems arising from the absorption of water by shales would appear to be the use of a drilling fluid that has oil as the liquid phase. Experience has shown, however, that oil muds always contain some water and that the hole stability sometimes is affected. Laboratory studies show that wet shales can be hardened by exposure to invert-emulsion mud that contains a high-salinity water in the emulsified phase.

Two methods have been used to estimate the salinity required. The first method (Mondshine) equates the surface hydration force of shale with the matrix stress (equal to the overburden pressure minus the pore fluid pressure). The salinity of the interstitial water is measured or estimated, and the required salinity is read from a graph. The other method (Chenevert) involves the measurement of the equilibrium vapor pressure of the shale (from cuttings) and adjustment of salinity of the emulsified water to the same or somewhat lower vapor pressure. As a practical field approach, the salinity of the water in the oil mud is raised to a concentration substantially above that estimated for the water in the shale. Maintaining a stable emulsion takes advantage of osmotic forces across the semi-permeable membrane to transfer water from the shale into the drilling fluid. In this way, the borehole wall may be made stronger.

The following advice can be given:

1. Increase pH to 10 with caustic soda prior to drilling anhydrite.
2. Treat-out excess calcium with soda ash.

3. Increase caustic soda to minimize the adverse effect of CO_2.

2.4.4 Packer Muds

Tubing and casing failures caused by corrosion have required expensive workover operations in deep wells. Oil-mud compositions have been developed specifically to prevent such problems. Stability requirements at high temperatures are severe. Settling of solids and extreme gelation or solidification must be avoided. Filtration control is not critical for mud to be left in the tubing-casing annulus but required for mud in the casing-hole annulus. The density of the mud is an important consideration.

An oil mud that has been used in drilling the hole can be treated to make an annulus pack, provided the original components have the necessary long-time stability at bottomhole temperature. An organophilic amine-clay reaction product is added to develop gelling properties. Supplemental filtration-control material is added for the casing-annulus pack. There are no problems involved in placement of the packs.

If the hole has been drilled with water mud, attention must be given to displacement by the oil pack. The water mud should be thinned as much as is practical, while avoiding the settling of barite. A portion of the oil mud is treated with sufficient organophilic clay to produce a high-yield point, or a premixed oil-mud spacer can be used, and this thick oil mud is pumped ahead of the packer mud. After the water mud and the spacer pill have returned to the surface, the packer mud can be recirculated and further treated if necessary.

2.5 Gas (Air) Drilling Fluid

The term *reduced-pressure drilling* has been applied to drilling with a circulating medium with a density less than that of water. This class of drilling fluids ranges from dry gas through mist, foam, "stiff foam", to aerated mud.

2.5.1 Advantages and Limitations

The principal benefit derived from air and aerated drilling fluids is the gain in penetration rate resulting from the lowered differential pressure. Weak formations can be drilled without loss of circulation. Producing formations are not damaged by invasion of the drilling fluid. Problems arise with dry-air drilling water-bearing strata are penetrated. Cuttings stick to the wet borehole and may plug the annulus. After a water-producing formation has been entered, the amount of water coming into the hole will control the drilling rate. If water-sensitive formations are exposed, hole problems will develop. Often the difficulties involved in "mudding-up" an air-drilled hole offset the savings made during that period of fast drilling.

2.5.2 Dry Gas—Dusting

Natural gas or combustion gas is sometimes used. If natural gas is used, care must be taken to avoid the formation of explosive mixture with air. Air is the gas generally used.

An annular velocity of 3,000 ft/min is often set as adequate to clean the hole of cuttings when drilling with dry air, although rates below and above this figure have been recommended. Air requirements for typical hole and pipe sizes are usually calculated on the basis of these assumptions: 3,000 ft/min velocity; air and cuttings form a homogeneous mixture that has the flow properties of a perfect gas; and the geothermal gradient is applicable for the temperature of the gas.

2.5.3 Mist

When formations that contain enough water to produce wet cuttings are penetrated, a mudring may form whose presence will be indicated by a rise in standpipe pressure. If there is no actual flow of water, drying agents, such as CMC, may allow a continuation of dusting. Chemical plugging agents are rarely successful. The usual approach is to introduce water or mud into the air stream as a mist. Practices vary, depending on the effect of water on the exposed formations. In mist drilling, the air lifts the cuttings but liquid wets the hole. Foaming agents, organic polymers, bentonite, or all together, may be introduced in the water. Corrosion protection becomes necessary.

2.5.4 Foam

As the quantity of water entering the hole increases or as removal of cuttings becomes difficult, the transition is made to foam drilling. Foam is the lifting medium; air furnishes the volume. *Quality of foam* is the fraction of air in the total volume; 0.96 quality means 4 parts liquid to 96 parts air. Foam quality and stability depend on the composition of the foaming agent, the composition of the injected water, and the composition of the fluids entering the well. Some foaming agents are not effective in salty water, or the foam is broken by traces of oil. Selection of the foaming agent may determine the success or failure of foam drilling.

Foam allows pressure to become stabilized by preventing the accumulation of wet cuttings in the hole. Lower velocities are required to lift cuttings than with dry air; therefore, smaller compressor capacity is needed. If only a little water is entering the hole from the formations, sufficient water is injected with the foaming agent to keep the cuttings from sticking together.

2.5.5 Stiff Foam or Gel Foam

The gel foam technique is an extension of foam drilling. The method affords a means of drilling poorly consolidated formations with a low-density drilling fluid in which hole-stabilizing materials can be included. The stronger film will support larger cuttings than simple foam will. When stiff foam is used, the annular velocity is from 100 to 200 ft/min; consequently, the compressor requirements are much less than in other forms of air drilling (200

ft^3/min, more or less, depending on the amount of fluid entering the hole from the formations).

Stiff foam was introduced as a means of removing cuttings from large-diameter holes drilled in loosely consolidated formations. The original composition consisted of bentonite, 10 to 15 lb/bbl; guar gum, 0.2 to 0.5 lb/bbl; soda ash, 1 lb/bbl; and a foaming agent, 1% by volume. Subsequent modifications include substitution of peptized bentonite for bentonite and guar gum and replacement of guar gum by other organic polymers. The gel-foam mixture is injected at such a rate as to furnish a return flow that has the consistency of shaving cream.

The usual surface hook-up foam drilling requires a controllable low-volume injection pump to introduce the foam-forming material into the air stream. Another method employs a foam generator and the introduction of preformed foam into the pipe. The composition of the foam mixture is selected to provide stability in the presence of salty water and oil. No bentonite is used.

Numerous factors are involved in fulfilling the basic requirements for successful foam drilling; that is, the maintenance of a continuous column of thick foam from the standpipe to the blooey line. Surface injection pressure, drillstring torque, and regularity and appearance of the foam at the blooey line are factors that are considered in adjusting the composition and the injection rate of the foam-forming mixture and air. Rate of penetration and size of the cuttings are also important.

2.6 Application of Nitrogen

During drilling operation when loss circulation occurs, especially when a light-weight mud is used, nitrogen can be injected into the mud system at the standpipe to lower hydrostatic pressure on bottom (see table 2.2).

Nitrogen has the advantages over other aerated systems because pressure, volume, and temperature can be controlled to a great degree of accuracy. Temperature could be one of the factors to be considered while working with an aerated system. Assuming that 300,000 scf of nitrogen would be required over a period of time to control hydrostatic pressure at depth and that the gas were to be pumped into the mud system at ambient temperature, whereas the wellbore temperature was 130 °F, we have a new volume factor:

$$V_2 = V_1 \times \frac{T_2}{T_1} = 300{,}000 + 15 \, \frac{130 + 460}{70 + 460} = 33{,}916 \text{ scf.}$$

At greater wellbore temperatures, the volume increases proportionally and can result in unloading the annulus. Tables developed by suppliers are designed for a geothermal gradient of 1.6 °F per 100 ft of depth.

The nitrogen line is connected to the standpipe during drilling operations and nitrogen is pumped at the required rate until the kelly joint is down. Before breaking the kelly joint, the nitrogen pumping is discounted and a safety mudcap is pumped to simulate a depth of approximately 1000 ft in the drillpipe to prevent unloading. Low solubility of nitrogen in the mud system gives ample time to make up a joint before nitrogen works its way back to the surface. A heavy mud in the 14-lb range may reduce the amount of safety mudcap required to make up a new joint.

Often during drilling operations a lost circulation zone is encountered at shallow to medium depths, while a high-pressure zone is being drilled at a much lower depth. In this type situation, nitrogen can be injected into the mud system again at the standpipe to lower the hydrostatic pressure at shallow depths, while at the same time maintaining a safe overburden at the greater depths.

Due to the pressure of circulation that compresses the nitrogen bubbles, the pressure gradient of the mud in the drillpipe is not affected to a great degree by the addition of nitrogen. A mud pressure gradient of 0.467 with a pump pressure of 2,000 psi, will only be reduced to 0.440 with the addition of 50 scf of nitrogen per barrel.

Only after passing through the bit jets and the pressure will the nitrogen bubbles expand and lower the pressure gradient

TABLE 2.2 WEIGHT OF FLUID WITH NITROGEN ADDED

Mud wt. (lb/gal)	N_2 (scf/bbl)	Depth (ft) 1000	2000	3000	4000	5000	6000	7000	8000	9000
9.0	10	7.72	8.16	8.36	8.48	8.56	8.61	8.65	8.68	8.71
	20	6.26	7.19	7.63	7.88	8.05	8.17	8.26	8.33	8.39
	50	3.04	4.58	5.53	6.13	6.55	6.85	7.08	7.26	7.41
10.0	10	8.69	9.15	9.35	9.47	9.55	9.60	9.64	8.68	9.70
	20	7.16	8.15	8.60	8.86	9.03	9.16	9.25	9.32	7.42
	50	3.67	5.41	6.43	7.07	7.50	7.81	8.04	8.23	8.38
11.0	10	9.66	10.13	10.34	10.46	10.54	10.60	10.64	10.67	10.70
	20	8.08	9.11	9.85	9.84	10.02	10.14	10.24	10.31	10.37
	50	4.29	6.25	7.33	7.99	8.44	8.76	9.00	9.19	9.35

drastically. Compare this to the data in table 2.2, where a 9.0-lb mud reduces from 7.41 at 9,000 ft to 6.55 at 5,000 ft as the pressure is decreased in the annulus.

A continuous injection of nitrogen presents no great problem as far as spillover is concerned at the wellhead, unless a large amount of nitrogen is pumped, e.g., 100 to 200 scf/bbl. The mud is still routed to the shale shaker where the nitrogen is dissipated to atmosphere. Degasser has not been required when 50 to 75 scf/bbl was pumped into the mud system.

A documented case in Oklahoma reports that severe mud loss was encountered at 6249 ft. The entire lost circulation material amounted to 25 lb/bbl, giving a mud weight of 8.4 lb and a 35-second viscosity, yet 750 bbl were lost to the formation and returns ceased. In an attempt to drill the additional 50 to 100 ft necessary to complete the well, the drillpipe got stuck.

It was confirmed that the pipe stuck at 5287 ft due to formation caving, with no differential sticking above the caving zone. Drilling mud was retreated with no circulation material lost, viscosity increased to 45 seconds and a loss of 12 cc of water. A thousand feet of drillpipe and the 6¼″ drillcollars were to be recovered by washover, and fishing without assurance of circulation could increase the risk that the 7⅝″ OD washpipe would get stuck in the same manner as the original one. The nitrogen pumps used in Oklahoma had a minimum stroke of 200 scf/min, whereby the maximum practical mud-pump stroke during the washover was 4 to 6 BPM. Gas concentration thus ranged from 30 to 50 scf/bbl, resulting in an apparent nitrified density of 7 lb/gal from the nonaerated 8.4 lb/gal and in lowering the hydrostatic pressure at 6253 ft from 2726 psi, which adequately maintained stability at the weak zone. A lower gas concentration would have been satisfactory could the two systems (nitrogen pumps and mud pumps) have been adjusted to lower the gas-pump rate. A total of 294,000 scf was used during the five-day fishing job, drilling 42 ft of hole, and conditioning for casing and logging. (Refer to table 2.3 for drilling mud recap form and table 2.4 for drilling mud report.)

2.7 Drilling Fluids Program

The following program is based upon available data and reflects the best judgment of those involved in it preparation. The on-site mud engineer will monitor the mud properties continuously and, if warranted, will recommend changes to the drilling supervisor.

Depth interval (m)	Mud wt. (lg/gal)	Specific gravity	Viscosity (sec)	Water loss (ML)	System
0– 308	8.5– 8.8	1.02–1.06	45	N.C.	Water
308–2130	8.5– 9.5	1.02–1.14	40–45	20	Brine
2130–3600	9.5–14.6	1.14–1.75	38–40	6	Brine
3600–4938	9.5–11.5	1.14–1.38	40	4	Brine

Recommendations:

a) For all logging raise viscosity to 60.
b) Keep solids to a minimum at all times, but especially in the 6900 to 11,650 ft interval. Loss of circulation is

TABLE 2.3 DRILLING MUD RECAP FORM

COMPANY ______________________ STATE ______________ COUNTY ________________ FIELD ________

WELL NAME AND NO. ________________________ LEGAL DESCRIPTION ______________________________

CONTRACTOR __________________ WAREHOUSE ___________ ENGINEER ______________ TOTAL DEPTH & DATE ________

SPUD DATE ___________ NUMBER OF DRILLING DAYS ________________ TYPE MUD SYSTEM ______________________

CASING PROGRAM		BITS		CORING & TESTING PROGRAM
DEPTH	SIZE	NUMBER	SIZE	
1. ________ ft.	________ in.	________	______ in.	________________
2. ________ ft.	________ in.	________	______ in.	________________
3. ________ ft.	________ in.	________	______ in.	________________
4. ________ ft.	________ in.	________	______ in.	________________

MAKE-UP WATER __________ Cl^- ppm __________ Ca^{++} p

SURFACE PIT VOLUME ________________ bbl

PUMP AND LINER SIZES ________________________

SPECIAL EQUIPMENT ________________________

DATE	DEPTH	WEIGHT ppg	VISCOSITY			YIELD POINT	GELS	pH	API Filtrate	pf	Chlorides	Calcium	Sulfates	SAND	SOLIDS	OIL	WATER	HTHP				Remc
			Funnel API sec	Apparent cp	Plastic cp	lb/100 sq ft	0/10 min		ml	ml	1000 ppm	ppm	ppm	% by volume	% by volume	% by volume	% by volume	500 psi ___ °F				Num

REMARKS: ____

TABLE 2.4 DRILLING MUD REPORT

OPTIONAL		DRILLING MUD REPORT NO.
		DATE ______ 19__ / DEPTH ______
		SPUD DATE ______ / PRESENT ACTIVITY
OPERATOR	CONTRACTOR	RIG NO.
REPORT FOR	REPORT FOR	SECTION, TOWNSHIP, RANGE

WELL NAME AND NO.	FIELD OR BLOCK NO.	COUNTY, PARISH OR OFF-SHORE AREA	STATE/PROVINCE

DRILLING ASSEMBLY			CASING	MUD VOLUME (BBL)		CIRCULATION DATA	
BIT SIZE	TYPE	JET SIZE	SURFACE SET @ FT	HOLE	PITS	PUMP SIZE X IN	ANNULAR VEL. (FT/MIN) DP______ DC______
DRILL PIPE SIZE	TYPE	LENGTH	INTERMEDIATE SET @ FT	TOTAL CIRCULATING VOLUME		PUMP MAKE, MODEL / ASSUMED EFF %	CIRCULATION PRESSURE (PSI)
DRILL PIPE SIZE	TYPE	LENGTH	INTERMEDIATE SET @ FT	IN STORAGE	WEIGHT	BBL/STK / STK/MIN	BOTTOMS UP (MIN)
DRILL COLLAR SIZE		LENGTH	PRODUCTION OR LINER SET @ FT	MUD TYPE		BBL/MIN / GAL/MIN	TOTAL CIRC. TIME (MIN)

	MUD PROPERTIES	
Sample From	☐ FL. ☐ PIT	☐ FL. ☐ PIT
Time Sample Taken		
Flowline Temperature °F		
Depth (ft)		
Weight ☐ (ppg) ☐ (lb/cu ft.) ☐ Sp.G		
Funnel Viscosity (sec/qt) API @ °F		
Plastic Viscosity cP @ °F		
Yield Point (lb/100 ft²)		
Gel Strength (lb/100 ft²) 10 sec/10 min	/	/
Filtrate API (cm³/30 min.)		
API HTHP Filtrate (cm³/30 min.) @ °F		
Cake Thickness (32nd in. API/HTHP)	/	/
Solids Content (% by Vol.) ☐ calculated ☐ retort		
Liquid Content (% by Vol.) Oil/Water	/	/
Sand Content (% by Vol.)		
Methylene Blue Capacity ☐ lb/bbl equiv. ☐ cm³/cm³ mud		
pH ☐ Strip ☐ Meter @ °F		
Alkalinity Mud (Pm)		
Alkalinity Filtrate (P_f/M_f)	/	/
Alternate Alkalinity Filtrate (P_1/P_2)	/	/
Chloride (mg/L)		
Total Hardness as Calcium (mg/L)		

MUD PROPERTY SPECIFICATIONS

WEIGHT	VISCOSITY	FILTRATE

RECOMMENDED TOUR TREATMENT

OPTIONAL

REMARKS

OPTIONAL

PRODUCT INVENTORY										
STARTING INVENTORY										
RECEIVED										
USED LAST 24 HR										
CLOSING INVENTORY										
COST LAST 24 HR										

EQUIPMENT

OPTIONAL

DAILY COST	CUMULATIVE COST

OPTIONAL

Source: Courtesy of the American Petroleum Institute from *Drilling Mud Report Form*, API RP 13G: recommended practice drilling mud report form, second edition, May 1982, American Petroleum Institute, Washington, D.C.

possible, therefore, all solids-control equipment will be operating at maximum efficiency to maintain a minimum mud weight.

c) Mud weight, viscosity, and water losses should be adjusted to maintain maximum penetration rate consistent with a clean, stable wellbore.
d) Brine polymer mud system with 4% KCL (potash) that uses Drispac and seamud should improve penetration rate up to 30% and with good solids-control equipment up to 40% over a conventional system.
e) Low viscosity with brine polymer mud will allow gas to break out at the surface. This system is very good for unstable shale drilling.
f) Be sure to have the following on the rig and in working condition: mud cleaner, desander, desilter, and shale shakers with good screens.

Figure 2.1 shows pressure versus depth with normal and overburden gradients to aid in the development of a mud program.

2.7.1 General Terms

Density	Matter measured as mass per unit volume expressed in pounds per gallon (lb/gal), pounds per square inch per 1000 ft of depth (psi/1000 ft), and pounds per cubic foot (lb/cu ft). Density is commonly referred to as "weight."
Viscosity	Commonly called the funnel viscosity. The marsh funnel viscosity is reported as the number of seconds required for one quart of given fluid to flow through the marsh funnel. In some areas, the quantity is 1,000 cc.

Water loss	Measure of the volume of fluid lost through filter media (usually filter paper) when drilling fluid is subjected to a differential pressure.
Gel strength	Pressure unit usually reported in lb/100 sq ft. It is the ability of a colloid to form gels, it is the measure of the same interparticle forces of a fluid as determined by the yield point, except that gel strength is measured under static conditions.
pH	Hydrogen-ion concentration, a measure of the acidity or alkalinity of a solution. The pH numbers range from 0 to 14 with 7 being neutral. The numbers below 7 indicate acidity, while those above 7 indicate alkalinity.
Resistivity	The electrical resistance offered to the passage of a current, expressed in ohm-meters; the reciprocal of conductivity. Freshwater drilling fluids are usually characterized by high resistivity, salt-water fluid by a low resistivity.
Abrasiveness	The ability to grind or wear away by friction.
Plastic viscosity	A measure of the internal resistance to fluid flow attributable to the amount, type, and size of solids present in a given fluid.
Yield value	Resistance to initial flow of electric charges on the surface of mud particles.
Spurt loss	Influx of flow into the formation before pore openings are bridged and filter cake formed.

2.7.2 Drilling Fluids

1. Caustic soda (sodium hydroxide NaOH), used for pH control in all water-based muds.
2. Soda ash (sodium carbonate Na_2CO_3), used as a chemical precipitant for calcium in water-based muds, especially calcium sulfate in low-pH muds.

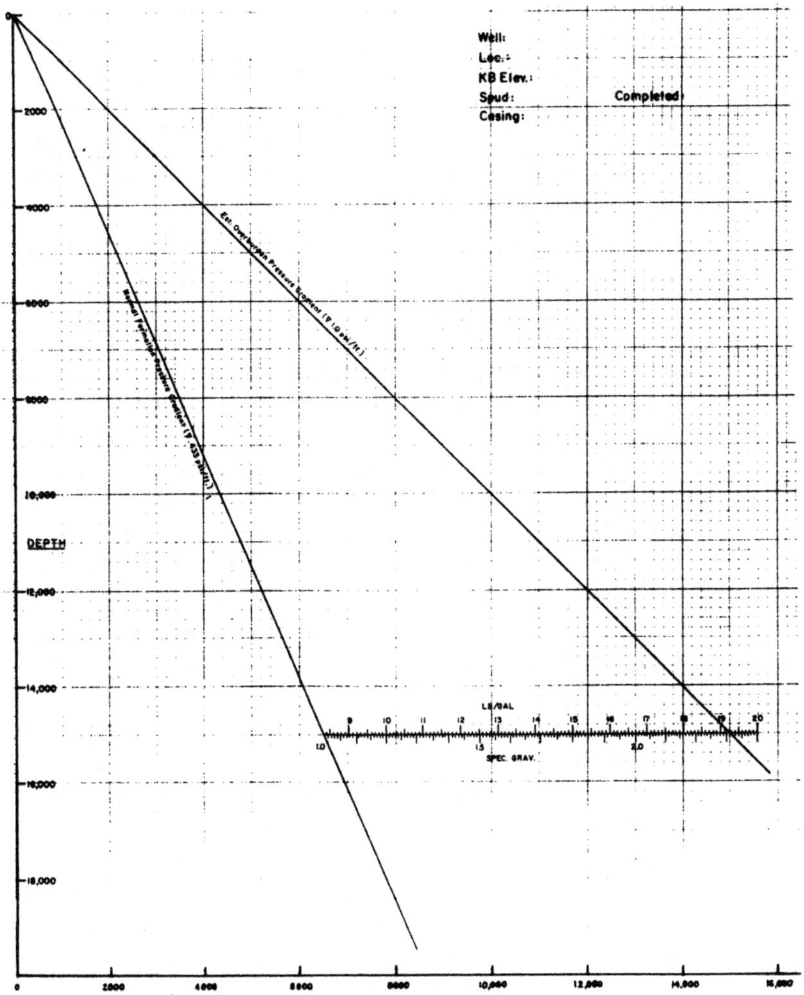

Figure 2.1 Shows pressure versus depth with normal and overburden gradients to aid in the development of a mud program.

3. Sodium bicarbonate ($NaHCO_3$), used for calcium precipitant (from cement of anahydrite) in muds above 8.5 pH.
4. Barium carbonate ($BaCO_3$), a chemical percipitant that may be used in low-pH muds where limited amounts of anhydrite are drilled.
5. Sodium chromate ($Na_2CrO_4 \cdot 10\ H_2O$), also sodium dichromate ($Na_2Cr_2O_7 \cdot 2H_2O$), used in all muds to prevent high-temperature gelation and to extend and rejuvenate the degradation of organic dispersants, thereby reducing the consumption of the dispersant. Also used as a corrosion inhibitor.
6. Gypsum (calcium sulfonate $CaSO_4 \cdot 2H_2O$), the source of calcium used to prepare gyp muds.
7. Lime (calcium hydroxide $Ca(OH)_2$), the source of calcium used to formulate lime muds. Lime may also be used to thicken mud for maximum hole cleaning.
8. Salt (sodium chloride NaCL), used in formulating saturated saltwater muds.
9. Caustic potash (potassium hydroxide KOH), sometimes used for pH stability and as an inhibitor for swelling shales.
10. Calcium chloride ($CaCl_2$), brine solutions for density range from 8.7 to 11.6 lb/gal.
11. Potassium chloride (potash KCl), used to control clay swelling.
12. Calcium chloride/calcium bromide ($CaCl_2$—$CaBr_2$), used for brine solutions to 15.1 lb/gal.
13. Zinc chloride ($ZnCl_2$), workover fluid to 14 lb/gal.
14. Calcium lignosulfonate (CLS), used as a thinner for freshwater, saltwater, and calcium-treated muds. Can contribute to fluid loss control and is nonpolluting.
15. Starch, and excellent fluid-loss reducer, particularly effective in salt and calcium muds.
16. Barite (barium sulfate $BaSO_4$), used to increase mud weights up to 20 lb/gal.
17. Bentonite (sodium montmorillonite), used to derive viscosity and fluid-loss control in freshwater muds. If

mud contains salt in excess of 35,000 ppm, it must be prehydrated before it is added to a salt-mud system.

18. Salt gel (attapulgite), used to derive viscosity in muds that contain salt in excess of 35,000 ppm. It does not control water loss and therefore starch or carboxmethyl cellulose has to be added for this purpose. Salt gel may be used to find spots where the attapulgite and added weight material has been left in zones of high porosity due to high rate of filtration.
19. Lime added to salt gel, seawater, and barite will increase viscosity.
20. Soda ash added to the above will reduce viscosity.

3
Drilling Problems

Problems can be encountered in the drilling of any hole at any time; and drilling personnel must be aware of this fact and of the symptoms that indicate the various types of problems. On the other hand, many problems can be avoided or minimized by proper planning. Geological information and the experience gained from drilling previous wells in the same area can be used to predict the existence of problem zones, and such data should certainly be used to the fullest extent. However, problems encountered in one well may not exist in the immediate vicinity, while new problems can arise at any time. Such is the nature of drilling a small hole to great depths through often heterogeneous layers of subsurface rock.

The major types of problems to be discussed here include loss of circulation, abnormal pressures and blowouts, sloughing formations, crooked holes, stuck drillpipe, and the control of formation fluids. Although discussed separately, many of these conditions are closely interrelated, and measures taken to prevent or ameliorate one may have an aggravating effect on others. Therefore, they must be considered as an ensemble both in well planning and during the actual drilling operation.

3.1 Loss of Circulation

Loss of circulation is probably the most common problem encountered in oilwell drilling; it can be caused by the character of the penetrated formation and/or by conditions induced during and by virtue of the drilling process. This condition can be defined as a significant and continuing loss of hole mud to a subsurface zone, so that there is only a partial return or no return of the annular mud stream to the surface.

There are, in general, four factors which cause the loss of circulation. First is the case of natural formation fractures which can occur in essentially any type of rock. If these fractures have sufficient width and density and, therefore, sufficient permeability and volume, no mudcake will be formed and hole mud can enter the fracture system in varying amounts. This type of loss is generally identified by a gradual lowering of the mud level in the pits. If drilling is continued into the fractured section, there is the danger of complete loss of returns. Remedial measures usually consist of adding the lost amount of circulation material to the mud, although sometimes a simple thickening of the mud will solve the problem.

Second is the case of induced fractures due to excessive mud weight or surging effect or too fast a downward movement of the drillstring. This situation can occur at any depth where zones of weakness or existing fracture systems are present. Commonly, weak formations above a zone of abnormally high pressure will break down under the forces created by actions taken to control the expected pressures. Such actions often consist of setting casing above the high-pressure interval and then drilling out with a heavy-weight mud. Weaker normal-pressure zones below the pipe are then exposed to the abnormally high mud pressure, often augmented by the pistoning effect of bit movement through the newly set pipe when making trips.

Loss of circulation to induced fractures is usually sudden and accompanied by a complete loss of returns. Anytime a loss of circulation occurs, and such has not been encountered in adjacent wells, induced fractures should be suspected. This is especially true when mud weight is 10.5 lb/gal or more and/or there has been any action that could have caused a sudden surge in pressure.

Remedial measures for this situation vary widely with the condition and its location in the hole. An obvious solution, in cases of fractures created in a long section of open hole, is to reduce the mud weight and add lost-circulation material. At times, pipe may have to be set or cement may have to be spotted in the zone before resuming normal drilling activities. Sometimes, the hole will have to cemented and side tracked. In any case, extreme caution must be taken, especially if abnormal pressures exist in the hole, because there is a serious danger of a blowout.

If the loss of circulation occurs in normal-pressure formations below the surface casing, the general solution is to drill with water without returns, letting the cuttings be deposited in the zone of lost circulation. Of course, this may require large volumes of water and the flow rate must be sufficient to lift and deposit the cuttings. There must be a close watch for any evidence of drill-string torque or drag. Occasionally, these problems can be solved by batch treatment with a lost-circulation material or a cement or diesel oil–bentonite plug.

A third case of lost circulation is that of cavernous formations, normally limestones. When such are encountered, the loss of returns may be sudden and complete and the bit may actually drop from a few inches to several feet, just before the loss of return is noticed. At times, the drilling may be "rough" for a few feet, prior to the loss of return. The usual remedy for this condition is to fill the cavernous section with some type of stable plug, such as cement. Crevices, channels, and large vugs may at times be sealed with batches of large-sized lost circulation materials or by batch treatments with cement or high-fluid-loss mud.

The fourth case of loss of circulation is that of unconsolidated formations with high permeability. These are usually shallow sands because permeabilities of the order of 50 to 100 darcy are usually necessary to cause the loss of hole mud. In this case, the drilling may be continued without circulation or the mud may be thickened to decrease the rate of loss, until sufficient depth is reached to set the surface casing. The danger here is that shallow washouts and caverns can be created, which can endanger the rig itself.

Finally, two additional observations are pertinent to this subject. A few years ago, the carrying of lost-circulation material in the drilling fluid as a precautionary measure was standard practice in many areas, and this is still done to some extent today. In practice, it has been found that, other than the cost, the use of fine lost-circulation materials in this manner does no harm in low-solids low-weight muds and can prevent seepage losses in low-pressure high-permeability zones. On the other hand, the use of course materials in low-weight muds is not a good practice, and the use of any type of precautionary lost-circulation materials in high-weight muds is a very poor practice, although it will still be encountered in some areas.

It should also be noted that care must be taken in pumping any lost-circulation material into a zone of induced fractures, because such materials can actualy serve as propping agents, thereby aggravating and prolonging the problem.

3.2 Abnormal Pressures and Blowouts

Blowouts and wells out of control have many causes, not the least of which are abnormally high formation pressures and situations resulting from actions taken for their control, such as those mentioned in the previous section. Statistically, nearly one-half of all land wells and more than one-third of all offshore wells drilled through abnormally pressured formations, experience a lot of trouble that leads to great expense, pollution of the natural environment, loss of petroleum reserves, and/or loss of human life. For this reason, an understanding of the origin, detection, and evaluation of abnormal pressure is important to anyone involved in the drilling of oil and gas wells.

By definition, abnormal pressures (or geopressures) are those greater than the normal hydrostatic pressure of the formation fluids. Pressure at depth can be expressed as:

$$P = 0.052\,pD,$$

where P = formation pressure, psi;
p = fluid density, lb/gal;
D = depth, ft.

In the region of the U.S. Gulf Coast, it is found that "normal" formation pressures are related to formation water with a salinity of approximately 80,000 ppm, or a density of 8.95 lb/gal. Therefore, for the Gulf Coast,

$$P = (0.052) \cdot (8.95) \cdot D = 0.465\,D,$$

or a normal pressure gradient of 0.465 psi/ft of depth. Anything greater than this would constitute an abnormal pressure (and anything less would be a subnormal pressure).

It has been observed that this "normal" gradient can and will vary from province to province and from a minimum of perhaps 0.426 psi/ft to a maximum of approximately 0.485 psi/ft. In "hard-rock" country, with older formations, the most widely accepted normal gradient is 0.429 psi/ft, or 8.25 lb/gal, which is less than that of fresh water.

In practice, abnormal pressures are described by a pseudo or false gradient. This is because the hydrostatic principle is violated in overpressured environments. Figure 3.1 illustrates the pressures and gradients for a typical Gulf-Coast well. Here, G_n is the normal gradient, and G_a represents the abnormal gradient at depth. In this case, the formation pressure is normal hydrostatic to 8,000 ft and abnormal below, reaching a maximum gradient of 0.85 psi/ft at 12,000 ft. Note that, unlike the normal gradient, the abnormal gradient varies with depth.

Generally, the maximum possible gradient is taken as 1.0 psi/ft, corresponding to a density of 19.33 lb/gal. The highest gradient ever substantiated was 1.04 psi/ft for a saltwater flow in the state of Mississippi, but the highest gradient recorded for any reservoir of consequence was 0.93 psi/ft. Several researchers have calculated from rock mechanics, that rocks under normal compaction cannot contain pore pressures greater than 0.96 psi/

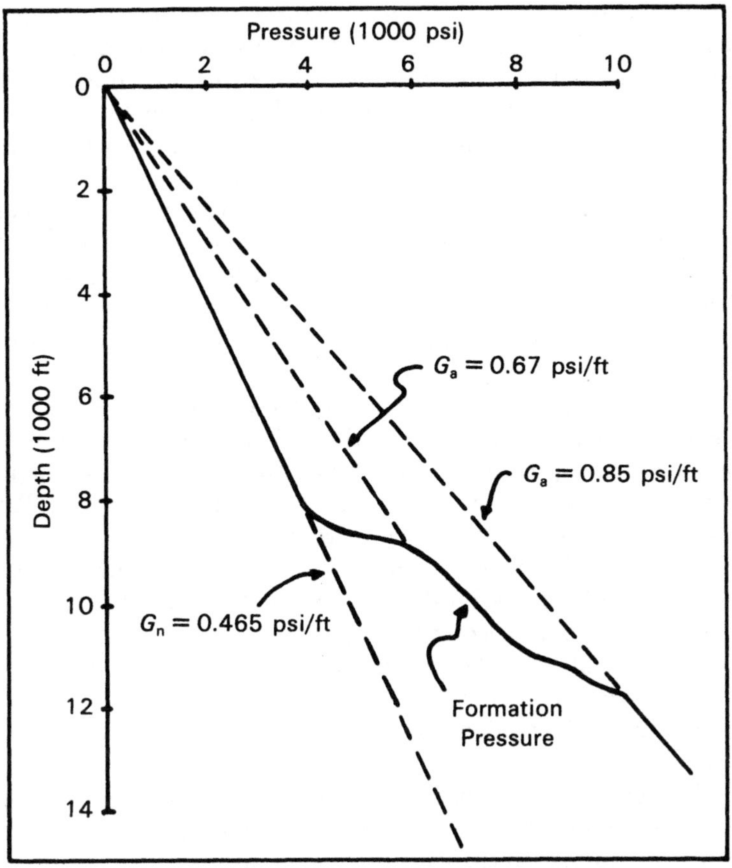

Figure 3.1 Pressure gradients in a typical Gulf-Coast well.

ft in the normal sequence of deposition, and this is equivalent to a fluid density of 18.4 lb/gal.

There have been many studies of the origin of abnormal pressures in rocks, and the best identifications suggest that such pressures are created primarily by four different phenomena: (1) hydrostatistically-pressured aquifers, (2) charged sands, (3) tectonic movement, and (4) compaction.

Frequently, relatively shallow aquifers will exhibit pressures well above normal because of structural conditions that cause them to outcrop at an elevation appreciably higher than that of the well-site surface. This phenomenon is common in mountainous areas and is the source of so-called artesian wells. Even though these formations show evidence of geopressures, the excess pressure is purely hydrostatic in nature.

High pressures can also occur in relatively shallow sands when they are charged by gas from deeper sands. This usually results from a poor surface casing cement job, casing leaks, or a blowout in a nearby well. Occasionally, upper sands can be overpressured if gas is trapped by very rapid deposition, but such occurrences are relatively uncommon.

On the other hand, geopressured environments are frequently caused by uplifting or faulting. If a formation is normally compacted at a great depth and than is lifted to a shallower depth without losing pressure, abnormally high pressure results. This phenomenon is common to many areas.

The geological process that uplifts the buried formation must also lift or fold the overburden. Therefore, the abnormal pressure exists only if some accompanying process reduces the distance from the formation to the surface; that is, simple uplift alone does not create abnormal pressures. The accompanying process can be piercement, deformation, erosion, or some combination of the three.

If subsurface movements are so severe as to cause faulting, deeper fluid pressures can escape to overpressured shallower formations similar to the case of charged sands.

Finally, compaction is the most common and perhaps best understood cause of abnormal pressures. Such pressures are the result of continued burial and compaction with restricted or blocked flow of fluids from the rock. This results in a transfer of overburden support from the rock matrix to the pore fluids, thereby causing pressures to approach that of the overburden gradient.

This process requires a seal for the rock to retain its fluids, and the presence and effectiveness of this sealing mechanism is the key to abnormal pressures by compaction. Therefore, the existence and magnitude of pressures in a particular sedimentary environment depend upon factors that impede the flow of formation fluids, including the presence and thickness of an impervious overlying formation, the depth of burial, the age of the sediments (the older the sediments the better the chance that pressures will drain off), faulting, absorption (liquid or gas molecules absorbed on shale surfaces can impede flow), and osmosis (semipermeable clay membranes can generate sufficient osmotic pressure to reduce permeability of the clay bed).

In addition to these four main causes, certain other processes can cause or contribute to abnormal pressures. Diagenesis in shales can release additional water to the pores and in carbonates can create permeability barriers. Secondary crystallization can increase pore pressure by causing a net increase in water mass per unit pore volume and can contribute to sealing. Simple thermal expansion of fluids can increase sealed pressures. Thermal cracking can increase hydrocarbon volumes two or threefold, thereby increasing pressure. All of these factors play at least a minor roll in most cases of abnormally pressured reservoirs.

The methods used for prediction and detection of over-pressured formations are many and varied. Some are direct and some are indirect, but generally they can be divided into three classes, depending upon the time applied; that is, before the hole is drilled, during the drilling of the hole, and when the well is tested and completed. These classes are summarized in table 3.1.

Much of the early work in the detection of overpressured zones demonstrated that if logged resistivity or sonic travel time is plotted versus depth, major distributions to the normal trends are attributable to abnormal pressures. As formations are compacted by increasing depth of burial and resulting overburden pressure, formation fluids are squeezed out and resistivity increases. Therefore, in normally pressured strata, resistivity will generally increase with depth. Overpressured zones, however, will have higher-than-normal porosity and fluid content with a resulting decrease in resistivity, as illustrated in figure 3.2.

The technique is used to plot pure shale resistivities on a logarithmic scale versus depth on a uniform scale. Best results are obtained where shale beds are at least 20 ft thick and where all available data are used. A trend line through these points should

TABLE 3.1 DETECTION AND EVALUATION OF ABNORMAL PRESSURES

I. PRIOR TO DRILLING
   1. Mud histories and drilling reports from offset wells
   2. Geologic correlations to similar areas
   3. Evaluation of offset wire-line logs
      a) Induction (conductivity)
      b) Electrical (resistivity)
      c) Acoustical (interval transit time)
      d) Gamma–gamma (density)
      e) Neutron–gamma (porosity)
   4. Geophysical surveys
      a) Seismic data (interval transit time)
      b) Gravity data (bulk density)

II. DURING DRILLING
   1. Kick
   2. Presence of contaminating formation fluids
   3. Increase in background and connection gas
   4. Abnormal trip fillup behavior
   5. Penetration rate/SP correlation
   6. Periodic log runs
   7. Paleontology
   8. Change in size and shape of shale cuttings
   9. Increase in fill on bottom
   10. Increase in drag and torque
   11. Increase in shale penetration rate
   12. Decrease in d-exponent trend
   13. Decrease in shale bulk density trend
   14. Increase in flowline temperature
   15. Increase in mud filtrate chloride content

III. AFTER DRILLING
   1. Drillstem tests
   2. Shut-in pressure tests
   3. Downhole pressure measurements
   4. Wire-line log evaluation

approach a straight-line increase with depth except where the shales are limey, are overlying a gas zone, or are heated due to the proximity of salt domes or massive salt beds.

For the same physical reasons, the acoustic travel time will increase in overpressured shales and will generally decrease with depth of burial. Therefore, anomalies on a semilogarithmic plot of transit time versus depth will often indicate the presence of overpressured zones, as shown in figure 3.3.

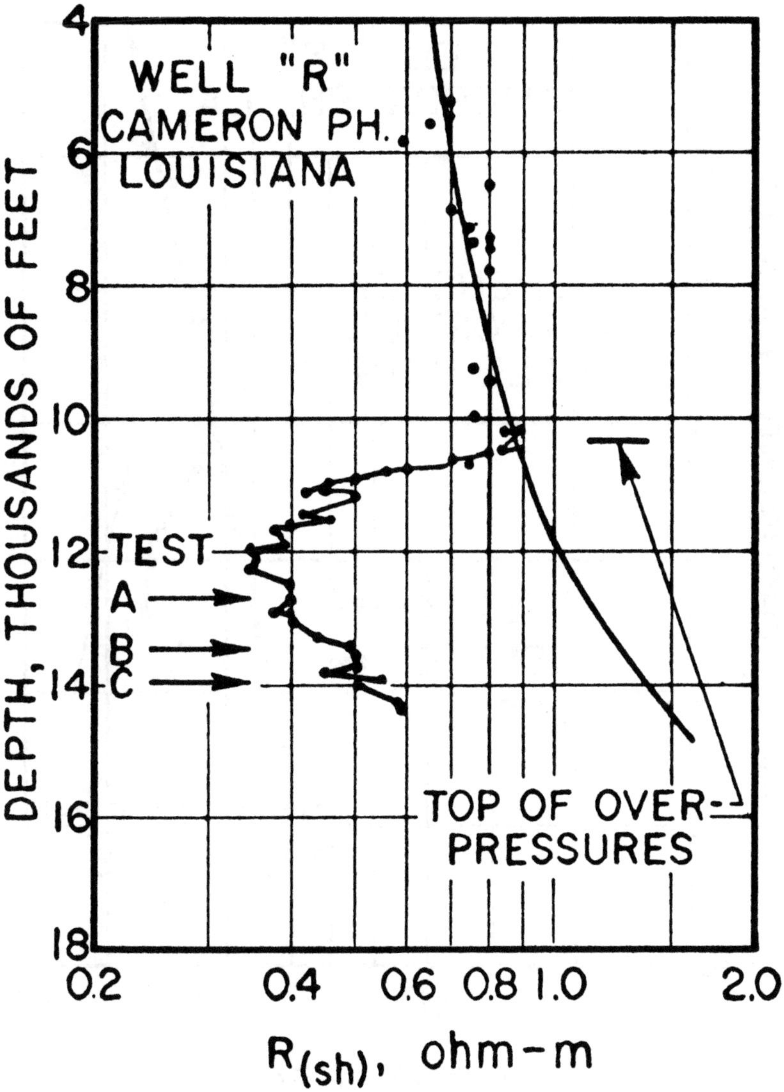

Figure 3.2 Plot of depth vs. resistivity.

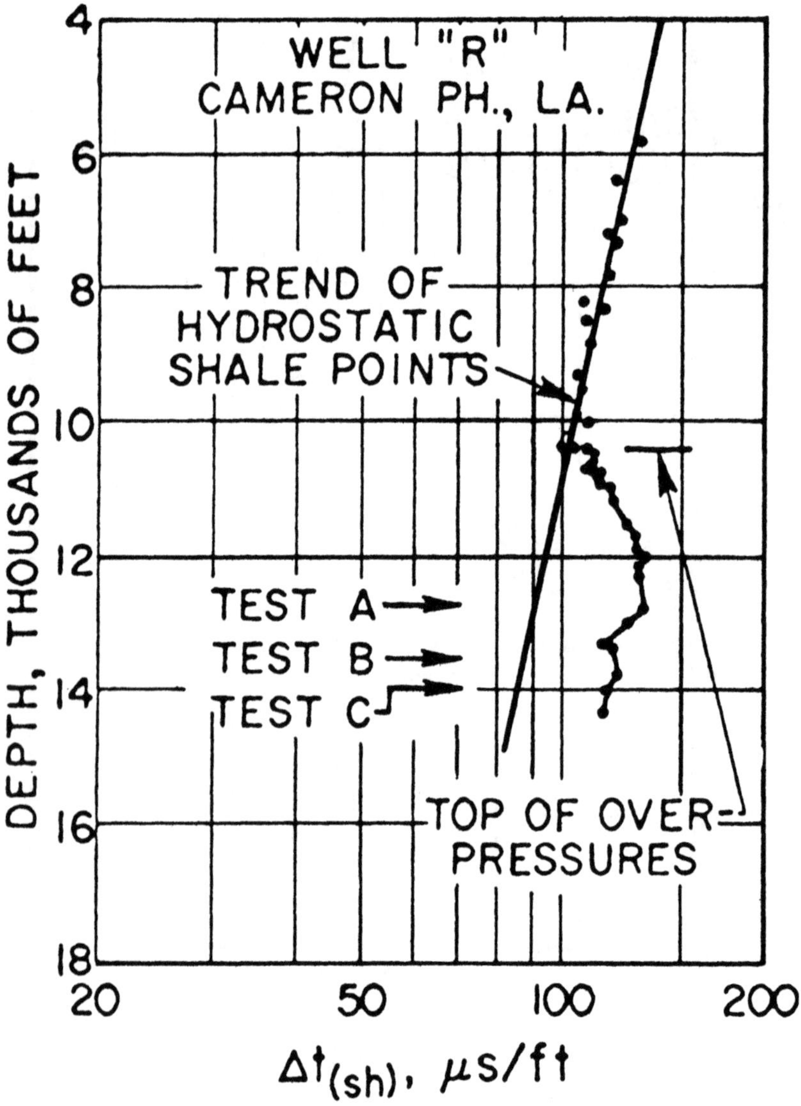

Figure 3.3 Transit time vs. depth.

Again, for the same physical reasons, shale densities and the size and shape of shale cuttings can be used to detect over-pressured zones. Since shale density generally increases with depth, decreasing density of shale cuttings will indicate abnormal pressures. By the same token, an examination of shale cuttings under a microscope often shows the cuttings from an over-pressured zone to be larger and more angular than normal. This is because, as the shale is drilled, excess pressure from the formation forces the fragments into the well bore and moves them more rapidly past the bit.

When shale density is the only measurable parameter available, it is useful to estimate pore pressure from a shale-density plot. Consider the case shown in fig. 3.4, where point x at the depth of interest has the same shale density as point y in the normally compacted section. If the overburden gradient and the normal pore pressure gradient are known, then the pore pressure at x can be calculated from:

$$P_x = (D_x) \cdot (\mathrm{OB}_x) - (\mathrm{D}_y) \cdot (\mathrm{OB}_y - \mathrm{PG}_y),$$

where:

P_x = pore pressure at x, psi,
PG_y = pressure gradient to y, psi/ft,
D = depth to X or Y, ft,
OB = overburden gradient to x or y, psi/ft.

If it is assumed that the overburden gradient is 1.0 psi/ft and that the normal pressure gradient is 0.465 psi/ft, then this equation becomes:

$$P_x = D_x - (0.535) \cdot D_y.$$

In any homogeneous formation, the temperature should increase linearly with depth under normal circumstances. Also, as shale becomes more compacted with depth, its pore fluid, which is effectively an insulator, is expelled and its thermal conductivity is increased. On the other hand, an overpressured shale, with its greater fluid content, forms a sort of partial thermal barrier, as illustrated in figure 3.5.

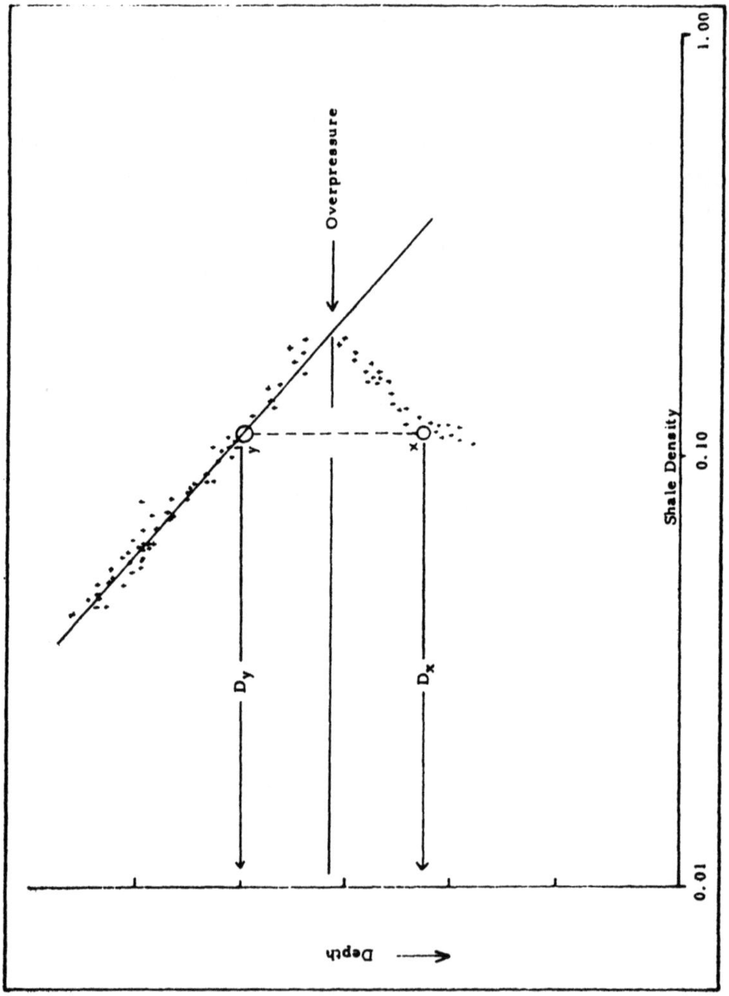

Figure 3.4 Pore pressure determination from shale density.

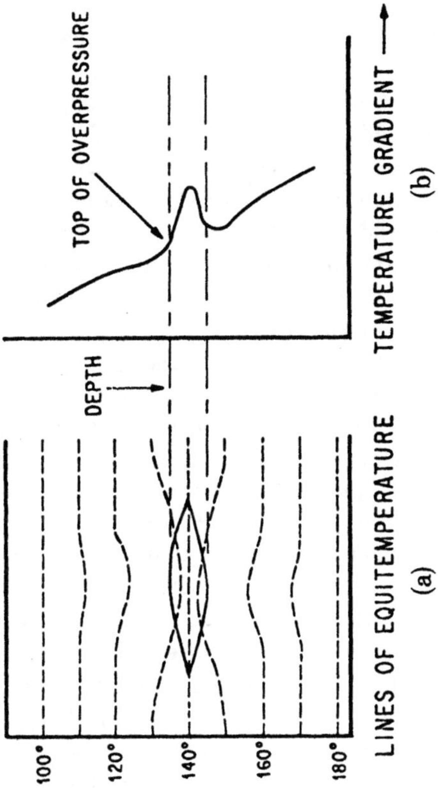

Figure 3.5 Indications of overpressure from flowline temperature.

In figure 3.5(a), the horizontal lens can be considered to be an overpressured zone within a continuous shale sequence, with the dashed lines being lines of equal temperature. Heat flowing out from the earth's core, encountering the overpressured zone, will build up below it until the thermal differences across it is great enough for it to be released. This heat buildup is represented by the upward distortion of the isotherms and by their wider spacing, which results in a reduced temperature–depth gradient.

Due to the buildup of heat below, the thermal gradient across the zone will be high. Then, as a result of the excess heat released through the zone, the isotherms are drawn down from above, and again, a reduction in the thermal gradient is noted. Detection of the overpressured zone depends upon recognition of this pattern of a reduced temperature gradient followed by a rapidly increasing gradient, as evidenced by monitoring the entering and returning mud temperatures.

Another means used for detection of overpressured zones consists of monitoring the chloride ion concentration by titration or resistivity measurements of the mud filtrate or the mud itself. Resistivity can be converted to equivalent chloride-ion concentration at the proper temperature (see fig. 3.6).

This method utilizes the premise that formation water becomes more saline with depth in normally pressured formations due to a charge imbalance on the montmorillonite clays, which causes a retention of dissolved salts as the formation water is squeezed out. In an overpressured zone, however, the porosity and water content are higher than normal, so that a trend reversal should take place, as illustrated in figure 3.7.

The *d*-exponent method utilizes a variable that has been used widely in distinguishing many different formation characteristics—from rock types to overpressured zones. It is a fundamental relationship of the rate of penetration.

Rate of penetration has been used, and used effectively, for many years. However, there is still a degree of uncertainty about the exact relationships between rate of penetration and the various controllable drilling variables.

It is known that drilling rate is a function of formation

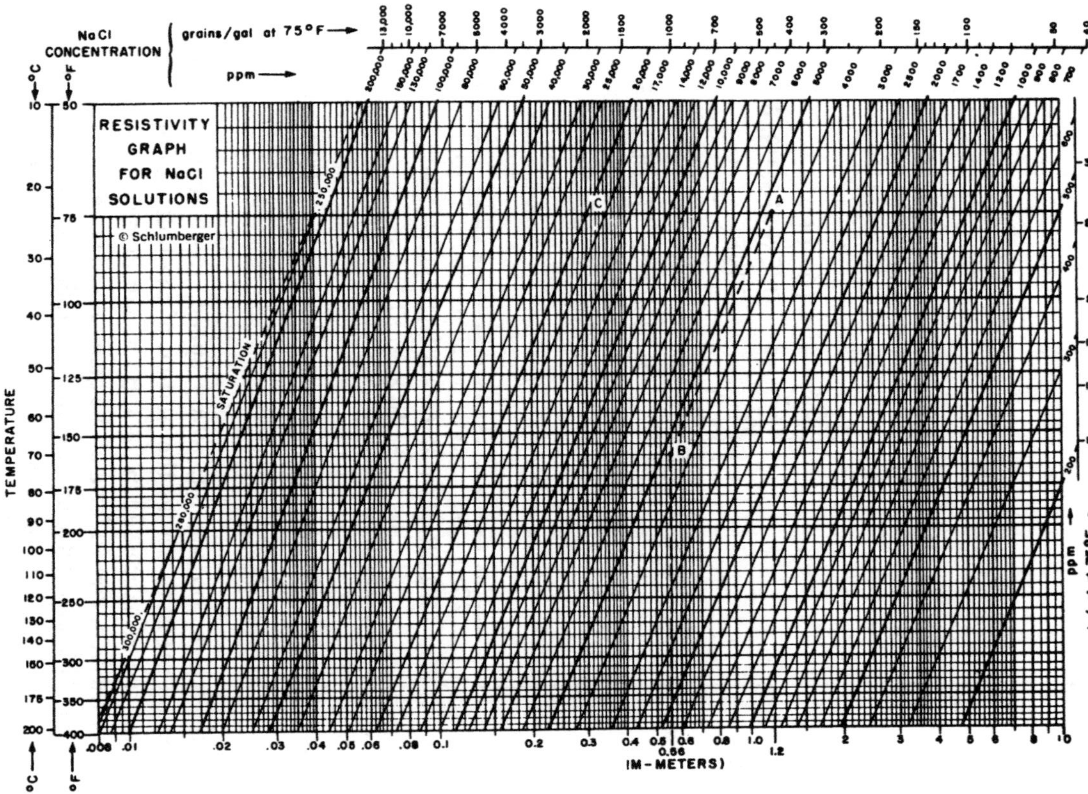

Figure 3.6 *Example:* R_m is 1.2 at 75°F (point A on chart). Follow trend of slanting lines (constant salinities) to find R_m at other temperatures; for example, at formation temperature (FT) = 160°F (point B) read $R_m = 0.56$. The conversion shown in this chart is approximated by the Arps formula: $R_{FT} = R_{75°} \times 75° + 7)/(FT(\text{in °F}) + 7)$. (Courtesy of Schlumberger Well Service)

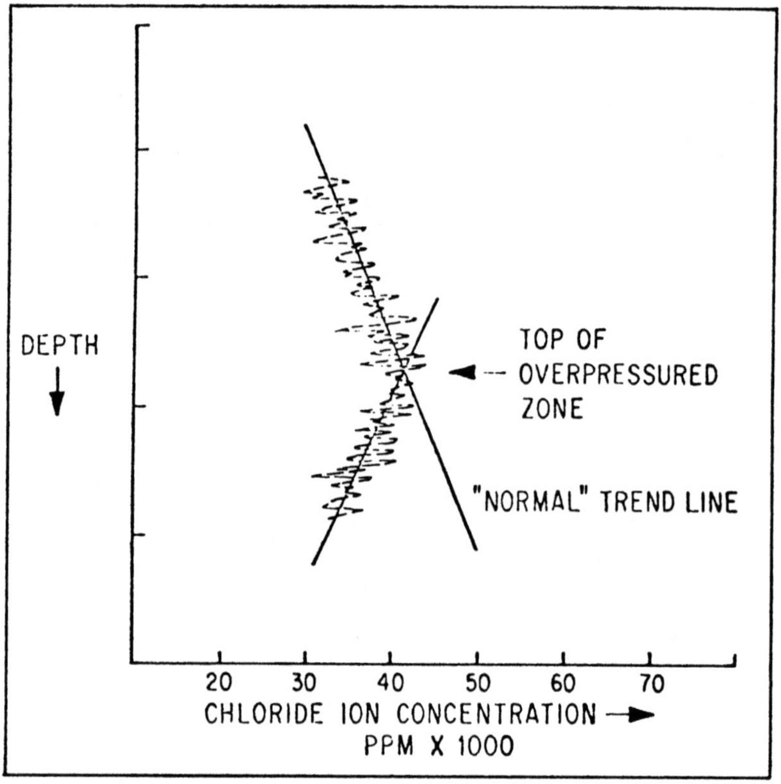

Figure 3.7 Indication of overpressure from chloride ion concentration.

strength, weight on the bit, rotary speed, tooth dullness and differential pressure across the bottom of the hole. All of these variables can be expressed in the form of a drilling rate equation. Possibly the best known form of the generalized equation is the *d*-exponent. This method is an attempt to stabilize or correct for the more significant drilling variables (specifically, weight-on-bit, rotary speed, bit diameter, and mud weight) in order for the penetration rate to reflect only changes in formation or formation pressure.

The equation of Jorden and Shirley relates penetration rate

to bit weight, rotary speed, and bit diameter when all other variables are constant:

$$\frac{R}{N} = k\left(\frac{W}{D}\right)^d,$$

where k = rock drillability constant,
D = diameter of bit, in.,
N = rotary speed, rpm,
W = bit weight, lb,
R = rate of penetration, ft/hr,
d = d-exponent.

This is a simplified form of the equation in that the effect of tooth wear and differential pressure are assumed to be unity; k is a dimensionless drillability or rock strength constant for each specific rock.

Therefore, this equation states that penetration rate per bit revolution is proportional to weight on the bit per inch of bit diameter, raised to some power d. Taking the log of both sides and solving for d, results in:

$$d = \frac{\log (R/N)}{\log (W/D)}.$$

By standardizing the units: $60 \times N$ to give rotations per *hour* and feet per *hour*, $12 \times W$ to give pounds per *foot* of bit diameter, and by assuming a set rock drillability of unity ($k = 1$), we can write the equation as:

$$d = \frac{\log (R/60N)}{\log (12W/10^6 D)},$$

where, 10^6 is a scaling factor to keep the ratio in a more manageable form.

By using a general drilling equation, such as the d-exponent, drilling performance data can be normalized and used in the detection of overpressured zones. A plot of *d*-exponent (on a logarithmic scale) against depth (on a linear scale) is constructed and, in normally pressured formations, a trend line of steadily increasing values with depth will develop. As the drilling parameters are normalized, the increasing *d*-exponent is a reflection of increasing formation competance with depth. In an overpressured zone, however, where porosity and pore pressure are abnormally high; penetration rates will increase and, correspondingly, the *d*-exponent will decrease—a reversal of the normal trend.

Representative plots of *d*-exponent and the corresponding formation fluid pressure gradient for two South Louisiana wells are shown in figure 3.8. Well A shows the idealized increase of *d*-exponent with depth under normal pressure followed by a reversal in the overpressured zone. The *d*-exponent plot for Well B is less distinct, but shows several trends of reversed slope in the overpressured zone.

A nomogram that can be used to determine the *d*-exponent is shown in figure 3.9.

Kicks and blowouts can be caused directly by the unexpected penetration of an abnormally pressured interval or by lost circulation resulting from, among other things, the methods used to control geopressures. In fact, circulated drill shows, cuts, and kicks are all related to mud properties and pore pressures. If these are not properly controlled, blowouts can and do occur. Definitions and the general rules to follow when such events are encountered are given below.

Drill show. This generally occurs when tight rocks are being drilled as the pore gas in the cuttings expands on the trip to the surface. Surface indications are:

1. gas bubbles in the mud and on the cuttings;
2. a cut in the mud weight of 1.0 lb/gas or less;
3. a duration of 30 min or less with quick dissipation.

Generally, no action is required.

Cut. This occurs when more gas is in the wellbore than can be attributed to the pore volume of the cuttings. This is usually

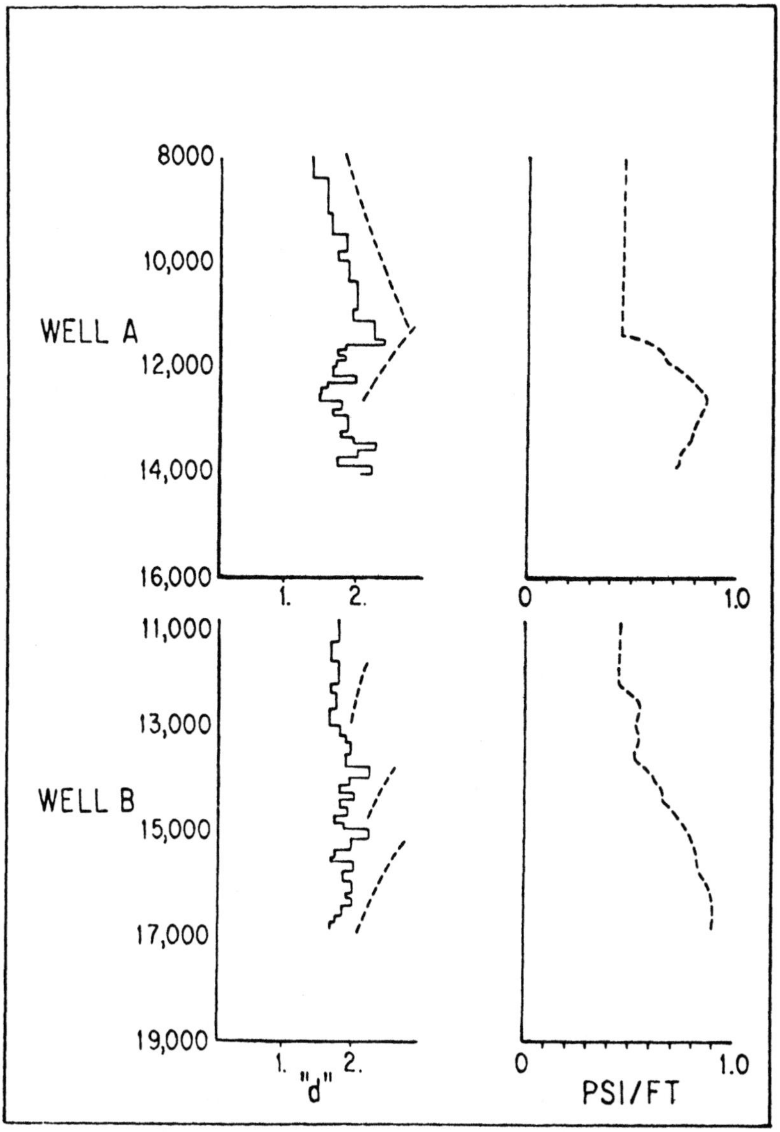

Figure 3.8 *d*-exponent and formation fluid pressure gradient.

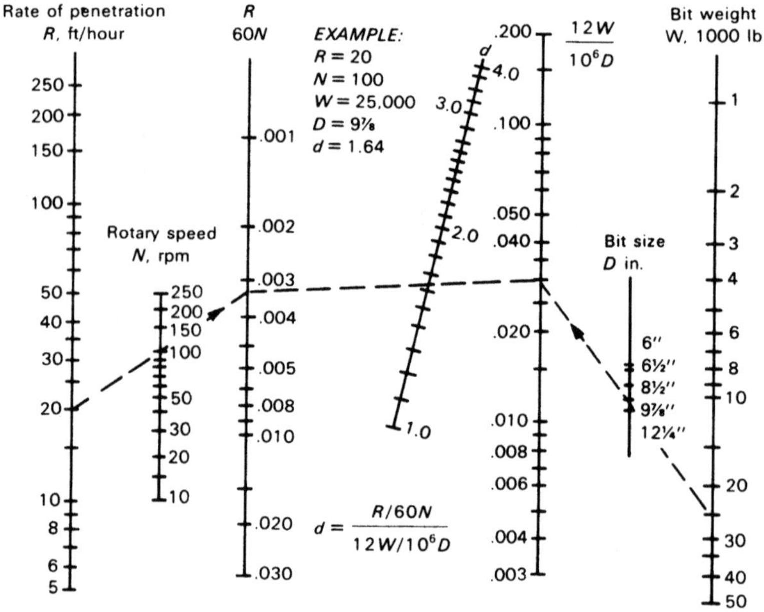

Figure 3.9 "d" exponent monogram.

encountered when a drilling break is being circulated up and after trips. Surface indications are:

1. a cut in the mud weight by 1.0 to 2.0 lb/gal;
2. a duration of 30 min to two hours.

This usually indicates that the formation pressure is just balanced. Again, no action is required, but the situation should be watched.

Kick. This is a situation where there is a continuous cutting of mud weight of more than 2.0 lb/gal, and there is a shut-in pressure on the drill pipe and/or casing. This is a dangerous situation that requires immediate action to kill the well and prevent a blowout.

A potential blowout is signalled by a sequence of events:

1. a drop in mud weight,
2. an increased flow at the flow line,
3. an increase in pit volume, and, finally,
4. the arrival of gas (or water) at the surface.

The general rules for dealing with a kick are:

1. Shut in the well kick as soon as possible.
2. Determine the shut-in drillpipe (and formation) pressure.
3. Circulate through a choke with a bottomhole pressure slightly higher than the formation pressure.
4. Increase the mud density before, during, or after circulating, as appropriate, to a weight that will contain the formation pressure and provide a margin of safety while pulling the drillpipe from the hole.

As always, the best remedy for kicks (and blowouts) is to plan and operate in a manner so as to prevent their occurrence. In this light, it is interesting to note that a study of 55 blowouts in California indicated the following causes, with the percentage of each shown in parentheses:

1. failure to keep the hole full on trips (42%)
2. lost circulation caused by fracturing a weak formation by too heavy a mud weight or by severe pressure surges when going into the hole (22%)
3. swab conditions when pulling out of the hole (16%)
4. insufficient mud weight (15%)
5. others, including allowing a mud column to gel and dehydrate while standing in a static condition (5%)

Tables 3.2 and 3.3 show the causes and remedies for blowout control. Table 3.4 lists some general rules of thumb for blowout control. Table 3.5 is a blowout control work sheet. It should be continually updated on the rig when drilling in a dangerous zone.

TABLE 3.2 BLOWOUT CONTROL

Occurrences	Warning signs	Causes	Action to take
When drilling	Increase in mudpit volume	Drilling fluid too light	Close B.O.P. and increase mud weight
When drilling	Loss of circulation	Drilling fluid too heavy or improper drilling techniques, i.e., spudding of drillpipe with pump on, etc.	Observe fluid level in annulus to insure that well is not kicking. If well kicks, close B.O.P. and add lost-circulation material to mud. Increase mud weight only if necessary.
When drilling	Gas, oil, or salt water cut mud	Swabbing or drilled show	Condition mud and increase mud weight only if showing was of sufficient intensity to warrant doing so.
When drilling	Increase in pump speed or decrease in pump pressure	Mud too light or hole in drill pipe	Observe mudpit level; if level increases, close B.O.P. and raise mud weight. If no increase in mudpit level, check for hole in drillpipe.
When making trip	Increase in mudpit volume or well beginning to flow back	Failure to fill hole and/or swabbing	Attempt to run drill pipe as near back to bottom as safety and time will permit. Close B.O.P., circulate and condition mud, increasing mud weight only if necessary.
When making trip	Hole fails to take mud when filling up	Swabbing	Do not pull any more pipe, but go back to bottom and condition mud. Lower gels and viscosity if high. Close B.O.P. and put on choke if necessary.
When making trip	Loss of circulation, unable to fill hole	Running drill pipe in hole too fast spudding of drill pipe	If unable to fill hole after waiting several hours, add lost-circulation material to mud system. Pay particular attention to mud in annulus to insure that well is not kicking.

When out of hole	Well begins to flow back, increase in mud-pit level	Failure to fill hole and/or swabbing	Attempt to run drillpipe as near back to bottom as safety and time permit. Circulate and condition mud. Close B.O.P. and put on choke, if necessary, and increase mud weight only if necessary.
When running casing	Loss of circulation	Running casing in hole too fast	Procedure to follow will depend upon the condition existing at time of losing circulation. In general, annulus should be kept full even to the extent of filling up with water in order to observe if well is kicking.
When running casing	Well begins to flow back	Swabbing on previous trip	If possible and if safety and time permit, continue to run casing to bottom. Close B.O.P. and put through choke, circulate and condition mud, raising mud weight only if necessary.
When testing	Well begins to flow back, increase in mudpit level	Completion fluid too light or well swabbed in when pulling DST tool or completion tool.	Kill well by pumping intact or by circulation and increasing weight of completion fluid.
Installing Christmas tree	Look around fittings	Improper installation and tightening of fittings, use of low-pressure fittings	First determine if fittings may be tightened or replaced, or if well must be killed. If in doubt, kill well. Do not take any chances.
After completing	Excessive pressure on annuluses	Leak in tubing or casing	Kill well and then make the necessary repairs.

TABLE 3.3 MECHANICAL AND HOLE PROBLEMS

Drillpipe pressure	Casing pressure	Action to take	Result	Problem	Solution
Up	Up about the same amount as drillpipe pressure	Check pump rate	Pump rate is too fast	Circulating pressure is too high because the pump is running faster than was planned	Slow the pump rate down to the planned rate. If pressures come down—everything is OK; if not—continue down chart.
		Increase choke size	Drillpipe pressure and casing pressure came down	Choke size was too small	If the pressures come down when the choke size was increased—everything is OK; if not—continue down chart.
		Open choke all the way	Drillpipe pressure and casing pressure came down	Either choke size was too small or the choke was trying to plug	If pressures come down—everything is OK; if not—continue down chart.
		Stop the pump	Drillpipe and casing pressure came down	The choke manifold has started to plug up	Switch to the alternate choke line and clear the manifold. If the pressures do not come down—continue down chart.

		Shut the well in	Pressures stay up	Manifold is plugged	Switch to alternate choke line, if pressures come down—go back to well killing, if not—continue down chart.
				Manifold is plugged at or above the "T"	Close the master valve on the kill line, release the pressure from the manifold and clean it out.
Up	Up, but not very high	Check pump rate	Pump rate too fast	Circulating pressure is too high because the pump rate is faster than planned	Slow the pump to the planned rate. If pressures come down—OK, if not—continue down chart.
		Increase choke size	Drillpipe and casing pressure came down	Choke size was too small	If the pressures come down—OK, if not—continue down chart.
			Casing pressure comes down, but not drillpipe pressure	WAIT 2 min to see if there is a long lag between choke movement and drill-pipe pressure	Allow for a long-time lag with big gas kicks. If pressure does not come down—continue down chart.
			Drillpipe pressure does not come down	A mud ring or packoff near the bit	Raise or reciprocate the drill pipe. If drillpipe pressure comes down—OK, if not—continue down chart.

TABLE 3.3 *(continued)*

Drillpipe pressure	Casing pressure	Action to take	Result	Problem	Solution
				Plugged jet	*Either*: Restore casing pressure to where it was before the trouble started. Take the changed drill pipe pressure as the new constant circulating pressure. *Or*: Stop the pump and shut the well in and bleed the pressure off the drill pipe. Then start up holding casing pressure constant until you reach a new pump rate. Then use the new circulating pressure as the new constant circulating pressure.
Up abrupt change	No change	Check pump rate	Pump rate too fast	Circulating pressure too high because rate is faster than planned	Slow the pump to the planned rate. If the pressure comes down—OK, if not—continue down chart.

		Increase choke size	Casing pressure gets very low before drillpipe pressure comes down	A mud ring or packoff near the bit	Raise or reciprocate the drillpipe. If drill-pipe pressure comes down—OK, if not—continue down chart.
Up abrupt change	No change	Increase choke size	Casing pressure gets very low before drillpipe pressure comes down	Plugged bit	*Either*: Take the new drillpipe pressure as the constant circulating pressure *Or*: Stop the pump and shut the well in and bleed off the drillpipe pressure. Then start up holding casing pressure constant, until you reach a new pump rate. Then use the new circulating pressure as the constant circulating pressure.
		Open choke	Drillpipe pressure does not come down	Plugged bit On a marine rig with subsea wellhead and riser a possible plugged wellhead or riser kill line	Stop the pump and shut the well in. Try "rocking" the pump to clear the bit. You may have to shoot-off or back-off the bit.

TABLE 3.3 *(continued)*

Drillpipe pressure	Casing pressure	Action to take	Result	Problem	Solution
No change	Down or no change	Increase or decrease in choke size	Pressures do not seem to respond to choke movement	Lost circulation, bad cement job, or a hole in the casing. Check pit volume	Pick a new slower circulating rate. Add lost-circulation material. Drop a barite plug.
		Check pit volume	Volume OK	Check the choke for failure	Switch to alternate choke.
Down	Down	Check pump rate	Pump rate too slow	Circulating pressure too low because the pump is running slower than was planned	Increase the pump rate to the planned rate. If pressures come up—OK, if not—continue down chart.
		Decrease choke size	Drillpipe and casing pressure came up	Choke size was too large	If pressures go up when choke size was decreased—OK if not—continue down chart.
			No change in drillpipe and casing pressure	Lost circulation, bad cement job, or a hole in the casing. Check pit volume	See: drillpipe pressure—no change.

Down	No change	Check pump rate	Pump rate too slow	Circulating pressure is too low because the pump is running slower than was planned	Increase the pump rate to the planned rate. If pressures come up—OK, if not—continue down chart.
		Decrease choke size	Pressures increase	Choke size was too large	If pressures go up when the choke size is decreased—OK, if not—continue down chart.
			Pressures increase but kelly hose jumps and drillpipe pressure surges	Pump trouble	Change pumps or repair pump.
		Continually decreasing choke size	Drillpipe pressure stays the same, casing pressure goes up	Hole in the drillpipe	Stop the pump and shut the well in. You may have to strip out to replace a joint of pipe.
Abrupt change down	No change	Decrease choke size	Drillpipe and casing pressure go up	Washout on bit or drill pipe	

(Courtesy of Dresser Industries, Houston)

TABLE 3.4 GENERAL RULES OF THUMB FOR BLOWOUT CONTROL

1. A jumping kelly hose or a surging pump pressure gauge are signs of pump problems.
2. If only the drillpipe pressure goes up abruptly, the bit is plugged or a nozzle is plugged.
3. If drillpipe and casing pressure go up abruptly, the choke or manifold is plugged.
4. If drillpipe pressure drifts down, look for a hole in the drillpipe.
5. If drillpipe and casing pressure don't seem to respond to the choke, check the pits for lost circulation.
6. When in doubt, stop the pumps, shut the well in, and **think**.

TABLE 3.5 BLOWOUT CONTROL WORK SHEET

1. Obtain before kick:

   Drillpipe capacity______________________________bbl/ft

   Pump output______________________________bbl/STK (stroke)

2. Obtain after shutting well in:

   Depth____________ft

   SIDPP (shut-in drillpipe pressure)__________________psi

   Mud wt.______________ppg

   SICP (shut-in casing pressure)______________________psi

3. Calculate formation pressure

   FP = 0.052 x Depth x MW + SIDPP.

   _________ = 0.052 x (_______) x (_______) + (________)

4. Calculated mud weight required

   $$MWR = \frac{19.2 \times FP}{Depth}:$$

   $$\_\_\_\_\_\_\_\_\_ = \frac{19.2 \times (\qquad\qquad)}{(\qquad\qquad)}$$

5. Calculate surface-to-bit strokes

   $$SBS = \frac{\text{DP capacity (bbl/ft)} \times \text{Depth}}{\text{Pump output (bbl/STK)}}:$$

TABLE 3.5 *(continued)*

______________ = ( ) x ( ) / ( )

6. Weight up mud to required weight.
7. Set pump on approximately 1/2 normal drilling stroke per minute and begin to circulate.
8. Using adjustable choke, maintain 50 to 100 psi over initial shut in casing pressure until new mud reaches bit.
9. Calculate surface-to-bit time

   SBT = SBS (min): / SPM

   __________ = ( ) / ( )

10. After new mud reaches to bottom, record standpipe pressure ______psi.
11. Using adjustable choke, maintain constant standpipe pressure until foreign fluids are circulated out.

3.3 Sloughing Shale

When formations tend to slough into the hole, the problem may or may not be sufficiently important to require or justify remedial measures. If such measures are needed and the condition is known to exist beforehand, then preventive measures are always preferable. Otherwise, when sloughing occurs, a decision must be made as to whether the problem can be ignored or some action must be taken to stop the sloughing and prevent further hole enlargement.

Unstable formations, which tend to swell and crumble into the hole, are almost universally called sloughing shales, but these include everything from clays, which are highly reactive to water, to completely lithified materials, such as claystones and slates, which are completely inert. In this context, the term is used in this all-inclusive sense.

Basically, sloughing shales cause two separate and distinct types of problems: hole fill, with the chance of drillpipe and test tools getting stuck, and hole enlargement, with the latter causing subsequent problems with cementing operations, especially if the sloughing formation is immediately between two productive zones that must subsequently be isolated one from the other behind the pipe. Another problem can arise when the sloughed material becomes a part of the drilling fluid, thereby altering the mud properties and increasing the cost of proper mud control. This is often a problem with sections containing larger amounts of hydratable clay, some of the worst offenders in terms of sloughing.

Sloughing and washouts can occur by drillpipe and drilling-fluid (turbulent flow) erosion, but most are associated with the effects of wellbore water on water-active clays. Therefore, montmorillonites are worse offenders than kaolinites, but studies have shown that many formations are capable of absorbing water or ions from water-base muds and water from oil-continuous muds. Such absorption occurs because of differences in chemical potential between the mud and the formation and takes place until equilibrium is achieved. With oil-continuous mud, only water can be absorbed because ions cannot move through the oil phase, but with water-base muds, the problem is more complex because of ion and clay movement as well as water movement between the mud and the formation.

Water and ion movement into "shales" is due to two separate pressure forces: the mud-to-formation differential pressure and the osmotic pressure across the interfaces. The first is a function of mud weight, depth, velocity, and pore pressure; and the second is a function of the chemical compositions and potentials of the mud and the formation.

Wellbores become unstable when formation pore pressure exceeds the pressure in the hole. Two situations can produce such a condition:

1. A normal wellbore pressure lower than the native pore pressure.

2. An on-site formation pressure increase due to water absorption.

Too low a wellbore pressure is the normal result of a too low mud density or the reduction in bottomhole pressure by swabbing with the drillpipe. This type of instability can usually be eliminated by drilling with an increased mud weight and/or by minimizing pressure surges with low-viscosity mud and slow pipe movement.

A more frequent cause of sloughing is the excessive formation pressure and volumetric swelling that occurs when argillaceous formations absorb water. These hydration problems usually occur some finite time after the formation has first been exposed and can be prevented or minimized in several ways to be discussed later.

Generally, deeper-seated shales present more of a hydration and sloughing problem than do the shallower zones. These more deeply buried shales are more compacted and contain lower quantities of water. This scarcity of water and the resulting abundance of charged-clay surfaces result in an increase in the absorptive potential of the shale.

From the previous discussion it should be obvious that a mud control and treating program can be used to prevent, reduce, or stop the action of sloughing formations. However, successful treatments usually depend on a change in mud chemistry rather than on a change in mud weight or filter-loss characteristics. In fact, decreasing the filtration rate is essentially useless because the tremendous osmotic pressures that can develop (of the order of 50,000 psi) will still result in wetting of the shale.

The chemical methods proposed and used through the years have been designed to alter the hydration characteristics of the clays and/or to isolate the unstable formations from the water phase of the mud, as discussed previously under the subject of drilling mud treatment. These methods have included the use of lime-based muds, special calcium-type muds, and oil-base muds. The newest method calls for the use of potassium chloride-polymer muds, which are claimed to prevent clay swelling by inhibition and encapsulation.

One technique, which was first introduced about 1966 and is

still applied in numerous cases, is to use a calcium–chloride solution as the water phase of an invert oil-base mud. A measured concentration of calcium salt can result in an ionic balance between the mud and the formation clay such that there is no net exchange of water and, hence, no swelling.

All of these methods involve some expense and, therefore, must be balanced against the choice of no treatment at all.

Shale Stabilization

The following steps can be suggested for shale stabilization:

1. Use inorganic salt to prevent hydration of the shale.
2. Use organic coating agent to prevent shale breakup.
3. Use a material for mechanical stabilization of the well-bore:
   a) potassium chloride (KCl) and ammonium sulfate inhibits shale, 10 to 35 lb/bbl
   b) flaxmeal—Drispac
   c) gilsonite—seamud

3.4 Deviated Hole

Some problems caused by deviated holes are:

1. completion problems
2. inadequate and misleading subsurface data
3. improper bottomhole locations with respect to physical and legal considerations
4. inefficient drainage of productive reservoir
5. excessive production and workover problems

From the standpoint of intentional (and correct) deviations, it has been proven that, with modern techniques and equipment, none of these problems need exist or are especially significant, even in bore holes inclined more than 50° from the vertical. Special techniques and equipment may be required in drilling, in running logs and pipe, and in interpreting subsurface data, but such are available. In fact, wells in the Huntington Beach field of California have been successfully deviated to as much as 80° from the vertical with no serious problems.

This fact is also true, from a mechanical standpoint, for unintentionally deviated holes. It has been observed that any drilled hole has a tendency to spiral, and some have made as many as three complete circles in 100 ft. It has been shown that, by the same token, the degree of spiraling decreases as the deviation from the vertical increases, with holes of more than 5° of deviation tending to move in a wide arc rather than in a true spiral. In the case of such gentle changes in direction, no remedial measures are necessary or justifiable except to force the hole into a required target circle at total depth. Otherwise, the cost and problems of corrective measures generally exceed those of simply drilling ahead.

The exception to this is the case of abrupt changes in direction which result in so-called dogleg. This situation can cause severe drilling problems and even endanger the hole. Problems that can be encountered, depending upon the severity of the dogleg, include:

1. fatigue failures of drillpipe
2. fatigue failures of drillcollar connections
3. worn tool joints and drillpipe
4. grooved casing
5. key seating of the drillpipe

Most of these problems greatly increase with the amount of tension in the drillpipe; therefore, the closer the dogleg is to the total anticipated depth, the greater becomes its acceptable severity because the lower end of the pipe has less tension stress than does the upper end. Figure 3.10 shows the influence of tension on the maximum tolerable dogleg angle for 4½-in. OD drillpipe.

A study by Arthur Lubinski, sponsored by the Crooked Hole

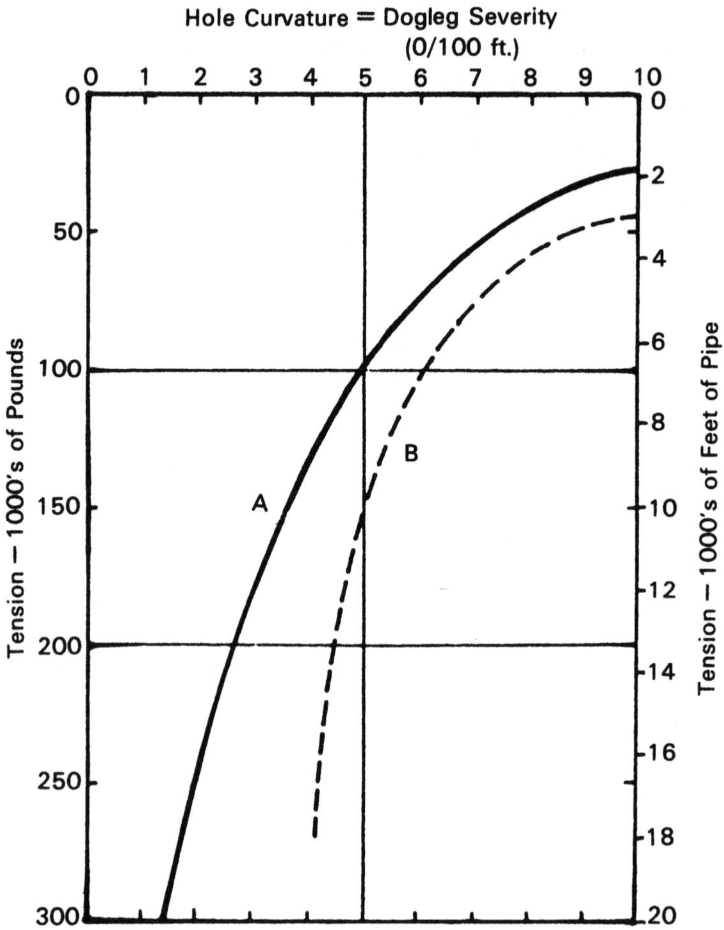

Figure 3.10 Hole curvature versus tension for 4½" drillpipe: *A*—fatigue of grade E drillpipe; *B*—drillpipe contacts wall. (For 5¾" OD tool joints.)

Sub-Committee of the IADC Rotary Drilling Committee and the API Mid-Continent Study Committee on Straight Hole Drilling, concluded that a dogleg is too severe if any of the following conditions exist:

1. The stress changes in the periphery of the drillpipe during rotation are sufficient to cause a fatigue failure.
2. The force in a dogleg on the side of the hole or casing is sufficient to cause wear of tool joints and drillpipe, to cause key seats in the wall of the hole, or to cause the tool joints to wear grooves in the casing.
3. The rotative stress reversals in the drillcollars are sufficient to cause fatigue of drillcollar connections.

Interestingly enough, this report and subsequent studies show that dogleg severity is not necessarily increased with increased weight on the bit and that rotating with the bit off bottom may be worse in terms of fatigue than drilling with the full weight of the drillcollars on the bit.

Preston Moore proposes that doglegging should be suspected when any of the following problems are encountered:

1. inability to log
2. inability or difficulty to run pipe
3. key seating
4. excessive casing wear
5. excessive drillpipe and collar wear
6. excessive drag forces
7. fatigue failures of drillpipe or collars
8. excessive wear on production equipment

The remedy for dogleg problems is to decrease the severity of the dogleg to a tolerable condition. This involves reducing the collar-to-wall clearance or packing the hole and increasing the stiffness by increasing the hole size. Available packed-hole equipment includes reamers, blade stabilizers, spiral collars, and rotating stabilizers. However, the most recommended procedure today is to use a square drillcollar, with a 1/16-in. corner-to-hole clearance, and optimum bit weights and rotary speeds. This technique has shown that few, if any, doglegs of more than 3½° per 100 ft will be created under any type of drilling conditions, even in so-called crooked-hole country.

Other procedures and practices that are used to cope with severe doglegs include increased frequency of drillstring inspections, reduced rotary speed, minimizing off-bottom rotation, packed holes, heavier casing, and string.

3.5 Stuck Drillpipe

Sticking of the drillpipe is still one of the commonly encountered problems in rotary drilling. Basically, this situation is caused by one of the three separate and distinct problems:

1. accumulation of cuttings or sloughed formation or balling of the bit
2. key seating
3. differential pressure

All three, mentioned in other sections, but discussed in more detail here, are illustrated schematically, in figure 3.11.

Pipe sticking due to the accumulation of materials in the hole is easy to visualize and is the result of inadequate removal of solids from the hole. Generally, this problem is not serious and can be remedied without too much effort. The problem is usually evidenced by increased circulation pressures and reduced or quenched free-fluid circulation.

A three-step procedure for remedying the situation calls for reestablishing circulation to remove the offending materials:

1. If circulation is not possible, shut off the pump, release the pressure, and work the pipe slowly. Then reestablish circulation with clear water, as in (2) below.
2. If circulation is possible, circulate clear water to remove the cuttings and debris.

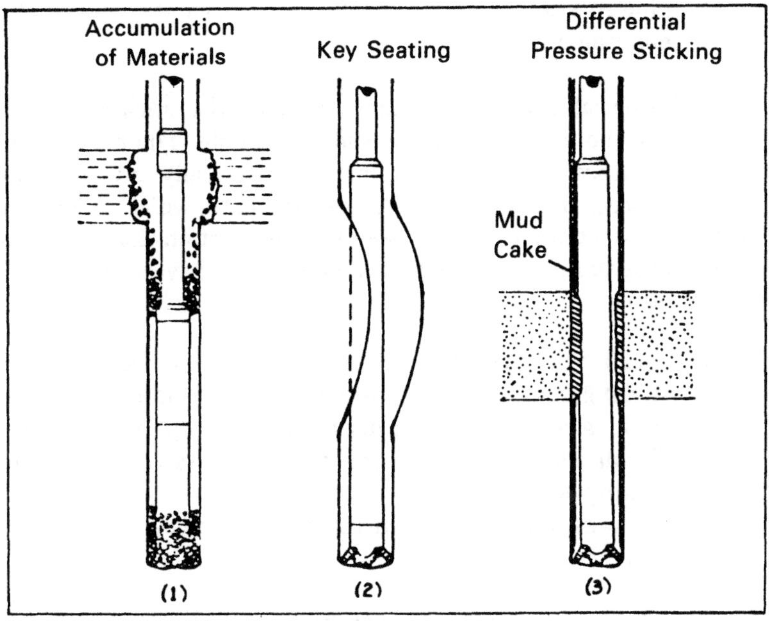

Figure 3.11 Three causes of a stuck drillpipe.

3. If pipe is still stuck, spot oil in the annulus to reduce friction.

Key seating is the result of severe doglegs and is usually formed at the shoulder where the hole straightens downward. At this point, a groove of the diameter of the tool joints wears into the side of the hole that supports the weight of the drillstring. This condition is evidenced by increased rotating torque to the point of stoppage or failure of the drill string. Circulation will still be free even though the pipe is stuck. As mentioned previously, the use of a square close-tolerance drillcollar will usually prevent this condition by eleminating severe doglegs. Methods for unsticking the pipe include:

1. a jarring action followed by rotation;
2. the spotting of oil to reduce friction; and

3. pipe rotation, if possible. Of course, circulation should be maintained.

Differential pressure sticking involves the sticking of the drillpipe in the mudcake on the wall of the hole. The mechanism of this process is quite complex, but basically involves the pressure forces from the hole *into* the formation, which causes the drillcollars to contact the mudcake and the friction that develops between the pipe and the clay of the mudcake. Several points are worth noting:

1. Sticking is generally restricted to the drillcollars.
2. Sticking occurs opposite a permeable formation.
3. Sticking occurs after an interruption of pipe movement.
4. Circulation, if interrupted, can be restarted after the sticking is noticed.
5. No large amounts of cuttings are circulated out after restarting circulation.

Also, because of changes that take place within the mudcake, both friction and pressure forces and, hence, the problem increase with time.

Conditions leading to differential-pressure sticking are:

1. oversize round drillcollars;
2. long strings of drillcollars;
3. high borehole deviation;
4. high weight, high water loss, high-solids mud, and
5. any interruption of pipe movement and long interruptions of circulation when mud quality is poor and annular velocity high.

Preventive measures include the avoidance of these conditions, coupled with:

1. reduced collar–cake contact surfaces by use of stabilizers, buttons, noncircular collars, etc., and
2. reduced friction between steel and clay by coating, oil muds, special mud additives, etc.

A remedial program that has proved satisfactory in many cases includes the following:

1. Stop circulation and apply maximum tension to drillstring as soon as possible.
2. If other conditions are suitable, reduce differential pressure by prolonged circulation of clear water or oil.
3. Spot oil and circulate in batches for prolonged intervals.
4. If all else fails, back off drillpipe and reduce pressure to atmospheric by connecting a formation tester to the collars. This method is often more successful if oil-base mud is spotted around the collars.

3.6 Control of Formation Fluids

The control of formation fluids, oil, water, and gas, is basically a problem of preventing unwanted-fluid entry by controlling hydraulic pressure in such a manner so as to offset the pore pressures in the formation while preventing loss of circulation or fracturing of the formations. The mud weight, fluid loss, and viscosity must be continuously monitored and adjusted, and pump pressures and rates must be set to maintain the desired subsurface mud characteristics and pressures.

In areas with abnormally pressured zones and/or with weak, incompetent formations, safe practices may involve setting several string of casing. Otherwise, formation fracturing and lost circulation may result in both weak and normally pressured zones with the mud pressures required to contain abnormal pressures. In normally pressured areas, weak zones may have to be protected even from the lower mud weights and pressures commonly used. In either case, it is necessary that the drilling team know as much about the area as possible and about the pore pressures and fracture gradients to be expected.

It is helpful also to find out as much as possible about the water sensitivity of potentially productive reservoirs. Sensitive formations, from either the standpoint of competency of subsequent production characteristics, may require special treatment in the form of decreased mud weights, chemical adjustment of the mud water, special waterless additives, and/or a changeover to an oil-continuous mud.

A special fluid entry problem encountered today is that of high-pressure high-rate capacity formations that produce gas with an appreciable hydrogen sulfide content. Drilling for such reserves and their subsequent production present a hazard not normally associated with the production of sour gas from moderately pressured reservoirs. This is the threat of blowouts caused by the often rapid failure of high-strength steels when they are exposed to hydrogen sulfide. When H_2S is present, limited-strength steels must be used and stringent quality control and inspection procedures must be established to insure that higher-strength materials are not used to drill, test, or complete the well. Contingency planning must exist so that surface equipment, rig personnel, and any nearby inhabitants and dwellings can be protected should a failure occur. Obviously, this becomes even more serious when wells are drilled in heavily populated areas.

Each of the various facets which are involved in the control of formation fluids have been discussed previously. Here, they all come together, and the student is urged to review the appropriate sections to form a comprehensive picture of the overall subject.

3.7 Bottomhole Assemblies

Good bottomhole assemblies, when properly designed and used, will:

1. Reduce rate of hole angle change. A smooth-walled hole with gradual angle change is more convenient to work through than one drilled at minimum hole angle with many ledges, offsets, and sharp angle change.
2. Improve bit performance and life by forcing the bit to rotate on the true axis about its design center, thus loading all cones equally.

3. Improve hole conditions for drilling, logging, and running casing; maximum size casing can be run to bottom.
4. Allow use of more drilling weight throughout formation that causes abnormal drift.
5. Maintain desired hole angle and course in directional drilling. In these controlled situations, high angles can be drilled with minimum danger of key seating or excessive pipe wear.

Bit stabilization has continued to receive more attention each year. Forty years ago, engineers wondered why 7⅞ in. bits performed better than 8¾ in. bits when they were both run with the same size drillcollars. The answer is that the 7⅞ in. bit was better stabilized than the 8¾ in.

Offset locations have been drilled in recent years in 60% of the time due to poor stabilization. It is suggested you use a stiff, stabilizing bottomhole assembly from top to bottom.

A recent development in stabilizers is a lock on tool, which permits its placement at any point in the drillcollar section. This makes it possible to get away from short drillcollars or placing the stabilizer at fixed points. This type stabilizer weighs up to 75% less, which is important from the handling and freight aspect.

A shock-sub is like a shock absorber on a car, in that it dampers the shock to the drillstring. This, in turn, gives longer life to tool joints, but it also gives longer life to the bit. This permits shorter drilling time for the well. It can be used as a jar for a limited period of time.

3.8 Drilling Practices

The high cost of drilling encourages everyone to improve the efficiency in drilling the well. Some steps in this direction are:

1. Explain to all personnel your drilling practices and procedures prior to commencing operations. Instill in them that the saving of any amount of time, no matter how small, is significant in an operation.

2. Thoroughly inspect, repair, and bolt down all pump discharges, manifolds, standpipe, and kelly hose.
3. Align rotary bushings and mouse hole to allow maximum speed in stabbing and making up pipe and kelly (connection).
4. Take deviation survey on dull bit and at a maximum of 1000 ft between surveys.
5. Do not ream kelly down prior to a connection or trip. Turn pump off and start out of hole as soon as kelly reaches rotary bushing.
6. Do not drill weight off bit prior to connection or trip.
7. Do not hit bottom with bit after connection or trip. Remember, weight was not drilled off and bottom will be encountered higher.
8. Do not circulate after getting on bottom after connection or trip. Turn pump on as soon as kelly is made up and start drilling immediately when bit hits bottom.
9. Do not break long tooth bits in before applying full drilling weight. Use short break-in time on hard formation bits.
10. Do not circulate prior to connection or trip.
11. Do not make a short trip at any time.
12. Run assigned pump pressures and linear size at all times.
13. Run bit nozzles as scheduled for various depths.
14. Place next bit to be used with correct nozzles on the floor prior to starting out of hole to change bits. Use the longest-tooth rotary bit possible until drilling rates and bit wear warrant a change to the next shorter tooth.
15. Run maximum weight on bit.
16. Turn rotary at the maximum permissible speed.
17. Train crews to make connections as rapidly as possible, probably, in 2 min or less.
18. Do all rig maintenance and greasing while drilling, if possible.
19. Instruct all crew members in blowout-prevention opening and closing techniques; however, the driller will perform this operation at all times, when practical.
20. Treat, convert, and weigh mud only while drilling.

21. If pipe becomes stuck, spot diesel oil immediately with the rig pump around the drill collars. Be rigged for this operation at all times.
22. If lost circulation is encountered, try to fill hole, but if it will not stand full, fill hole with water and measure the volume to determine the maximum mud weight the formation will stand. Pull up into casing as soon as possible.
23. If well starts to flow or kicks, shut well in and read the gauge on the standpipe to determine mud weight required to hold formation pressure.
24. Keep an inside blowout preventer on the rig floor for stabbing in the drill string at all times.
25. Keep a complete drawing, with dimensions of the drilling string, at all times.

3.9 Casing Design

In accordance with sound engineering principles and economy, casing must be designed so that it will not:

1. fail under tension
2. fail by collapse from external pressure
3. fail by bursting from internal pressure

Properly designed combination casing strings result in savings in casing cost and also improved tension factors of safety due to reduction in total weight of the string. Although combination strings consisting of many different weights and grades of casing can be designed, it is general practice to set some reasonable limit on the number of sections in the string.

Combination casing strings are designed to take advantage of the fact that the principal forces acting on a casing string are lowered in magnitude in the middle section than at the ends. Consequently, in this middle section, it is possible to use casing of lighter weight or lower strength (grade) than necessary in the top and bottom sections without reduction in the factors of safety for the string as a whole.

3.10 Cementing Operations

After planning the cementing program for a specific casing job and after designing the slurry, the actual procedure for placement of the cement downhole is the major consideration for a successful primary cement job.

3.11 Preparation and Running Casing

1. Mill varnish and scale should be removed from casing that will be set opposite the pay zone or other intervals that require absolute bonding.
2. Float equipment, scratchers, and centralizers should be installed while casing is on the rack, not while it is being run.
   a) Purpose of centralizers:
      to assure even distribution of cement around casing
      to control the annular spacing
      to improve wellbore cleanup and placement of cement
      to reduce frictional drag of casing
      to reduce differential pressure sticking
   b) Purposes of scratchers:
      to clean formation
      to clean out washouts and hole irregularities
      to control annular spacing
      to improve placement of cement
      to reinforce cement
3. The circulating head should be installed on a joint of casing which is placed in a readily accessible area.
4. If the string is very long, it is desirable to stop and

circulate the well at invervals. In deep wells, stop and break circulation before running below the previous casing shoe.

3.12 After Reaching Bottom

1. Start circulation slowly and begin reciprocating the pipe with short movements. Increase the pump rate and pipe stroke as pressure decreases. The stroke should be of sufficient length to cause overlap of scratchers, or up to 40 ft.
2. Circulation should continue as long as the shale-shaker screen shows return of mudcake and cuttings. Pump pressure and weight indicators also indicate when the hole is clean.

3.13 Conditioning the Casing and Hole

1. A chemical preflush or water should be pumped ahead of the cement to help remove the mud and filter cake from the casing and wellbore. This may consist of water and a surfactant or of a weak-acid solution.
2. If oil-base mud or inverted-emulsion muds are used, a surfactant and a dispersing agent are required to alter the wettability of the pipe and the wellbore.
3. Following the preflush, a scavenger slurry of 50 to 100 sacks of thin slurry ahead of the cement will further assist in cleaning the pipe and wellbore.
4. Continued reciprocation of the casing during this operation enhances the cleaning operation.

3.14 Mixing and Displacing the Cement

1. Two plugs are recommended to insure against contamination of the cement.
2. After releasing the bottom wiper plug, the cement is mixed and pumped at the predetermined annular velocity. At this point, pump pressure will be a minimum, and turbulent flow is easily obtained.
3. When all cement is mixed and is in the casing, the top plug is released, a bypass is opened, and cement is purged from the lines with water or mud. Water should not be pumped behind the cement before release of the top plug.
4. As the cement passes around the shoe, an increase in pressure will be noted, and the hanging weight of the casing will decrease. Slow reciprocal movement of the casing should be continued.
5. The displacing volume should be measured to avoid circulating mud around the shoe if the top plug fails.

3.15 Postplug Procedure

1. After the plug is bumped and the casing pressure is tested to the desired pressure, the casing reciprocation is continued to distribute excess water evenly. If increased drag is noted, the casing should be landed immediately.
2. When the casing is landed, the bleed-back line is opened to determine if the check valve in the shoe is holding. If it is, bleed all pressure off the casing, slack off the desired weight, and set the slips. *Note*: A careful watch should be maintained on the bleedline to prevent backflow if the check valve fails. This also applies to the valve on the casing houseing. If fluid is entering, this will cause the annulus to overflow.

3.16 Cementing

The reasons for cementing casing are well known. However, the most important ones are reviewed below:

1. control the well during further drilling
2. strengthen and protect all casing strings
3. support vertical and radial loading on casing
4. isolate porous formations
5. facilitate well treatment

The major reasons for cementing failures are:

1. cement channeling through mud
2. flash setting and/or loss of cement
3. damage by drillstring or perforating
4. movement of plastic salt or shales
5. migrating of water and/or gas through setting cement
6. no bond to formation and/or casing

Physical properties giving optimum viscosity, thickening time, and strength are of primary importance in designing cementing programs for wells. The optimum water-to-cement ratio is of utmost importance.

Displacement of the drilling mud from the annuli and replacement of the mud by cement, so as to fully encase the pipe in a uniform sheath of cement, will determine the final success of the job.

Turbulent flow is always considered desirable for cementations. Addition of friction reducers to promote turbulent flow can result in adverse viscosity relationships between the mud and slurry, which can result in channeling and increased thickening time and reduced strength of the cement.

Predicting flow behavior and fluid displacement in eccentric annuli when fully turbulent flow develops is difficult, however, velocity distribution around the annulus is less irregular. The flow favors the widest part of the annulus, therefore, displacement of a fluid from the narrow part of the annulus is not assured by turbulent flow, thus creating the channeling effect.

If the circulating fluid is a newtonian fluid, some flow will always occur in the narrow sector of the annulus. Otherwise, the yield strength of the fluid may prevent flow in this narrow region until the flow in the wider region is considerably higher. This means that it may be impossible to remove all the mud unless the casing is almost perfectly centered in the wellbore. Even then, if there are washouts, some gelled mud may remain in these washouts. This could occur because the cement flow in the enlarged area of the washout will revert to laminar or plug flow.

When heavy wall cake is built up on porous permeable zones and this cake is not removed prior to cement displacing, several conditions may develop:

1. The cake is not removed, resulting in poor cement bonding.
2. Being removed by the cement and transported up the hole, the mudcake bridges, allowing cement to bypass it.
3. The cake mixes with the cement and contaminates the slurry so that no set is achieved or cement strength is considerably weakened.

It has been demonstrated through extensive laboratory study that displacement efficiency can be significantly improved when the weight of the cement (displacing fluid) is equal to, or greater than, that of the mud (displaced fluid). These investigations also showed that if the weight of the two fluids were equal, but the yield point of the displacing fluid (cement) was greater than that of the displaced fluid (mud), the displacement efficiency was greatly improved. Similar results were obtained when yield points were equal and the cement weight was greater than the mud weight.

To improve the chances of getting a good cement job, do the following:

1. Circulate and condition the mud until the best possible gel strength and weight are realized prior to running casing.
2. Pump a good preflush (acid and dispersant) to remove mud cake and gelled mud. Careful attention should be given to the hydrostatic head on the annulus when calculating preflush volume.

3. Cement slurry design should provide for a maximum differential between cement and mud weight and yield point.

If a highly fractured zone is encountered in the cementing area, the Weatherford cleavage barrier should be considered as a method to improve the quality of cement over this interval. This technique spaces the scratcher one foot apart through the interval. The scratcher performs the extra function of reinforcing the cement sheath in addition to cleaning the wellbore and pipe during the cementing operation. This procedure helps to minimize cement damage during perforating.

3.17 Casing Selection Chart

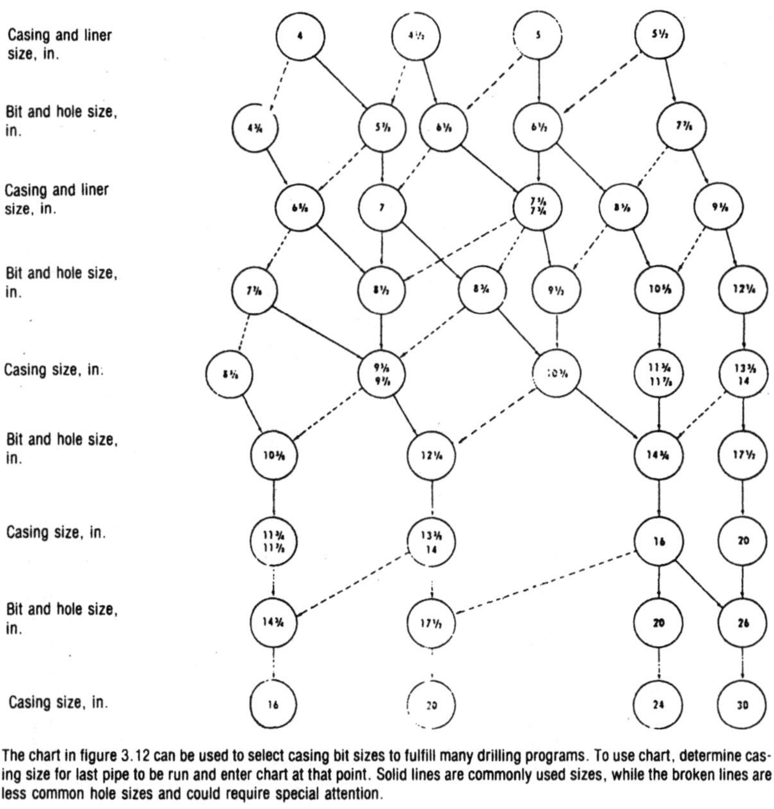

The chart in figure 3.12 can be used to select casing bit sizes to fulfill many drilling programs. To use chart, determine casing size for last pipe to be run and enter chart at that point. Solid lines are commonly used sizes, while the broken lines are less common hole sizes and could require special attention.

Figure 3.12 Casing selection chart.

Table 3.6 lists the details of sizes, weights, and clearances to be used after using figure 3.12 to establish the desired hole and casing design. Table 3.7 is a casing job work form to assist in performing as trouble free a casing and cementing job as possible. Table 3.8 is a liner job form. It can be used to help insure good performance in running, setting, and cementing a liner. Table 3.9 is a casing tally summary sheet that can sometimes be used instead of all the pages of the actual field tally.

TABLE 3.6 API CASING DIMENSIONS & BIT CLEARANCE DATA

Dimensions						Bit Size and Diametral Clearance		
Size OD	Weight Per Foot Nominal	Wall	Nominal ID	Coupling OD	Drift	Nominal Bit Size		Clearance From Drift Dia.
Inches	Pounds	Inches				Inches	Decimal	Inches
	9.50	.205	4.090	5.000	3.965	3⅞	3.875	.090
4½	11.60	.250	4.000	5.000	3.875	3⅞	3.875	.000
	13.50	.290	3.920	5.000	3.795	3¾	3.750	.045
	15.10	.337	3.826	5.000	3.701	3⅝	3.625	.076
	11.50	.220	4.560	5.563	4.435	4¼	4.250	.185
5	13.00	.253	4.494	5.563	4.369	4¼	4.250	.119
	15.00	.296	4.408	5.563	4.283	4¼	4.250	.033
	18.00	.362	4.276	5.563	4.151	4⅛	4.125	.026
	13.00	.228	5.044	6.050	4.919	4¾	4.750	.169
	14.00	.244	5.012	6.050	4.887	4¾	4.750	.137
5½	15.50	.275	4.950	6.050	4.825	4¾	4.750	.075
	17.00	.304	4.892	6.050	4.767	4¾	4.750	.017
	20.00	.361	4.778	6.050	4.653	4⅝	4.625	.028
	23.00	.415	4.670	6.050	4.545	4½	4.500	.045
	15.00	.238	5.524	6.625	5.399	5⅜	5.375	.024
	18.00	.288	5.424	6.625	5.299	5⅛	5.125	.174
6	20.00	.324	5.352	6.625	5.227	5⅛	5.125	.102
	23.00	.380	5.240	6.625	5.115	4⅞	4.875	.240
	26.00	.434	5.132	6.625	5.007	4⅞	4.875	.132
	17.00	.245	6.135	7.390	6.010	6	6.000	.010
	20.00	.288	6.049	7.390	5.924	5⅞	5.875	.049
6⅝	24.00	.352	5.921	7.390	5.796	5¾	5.750	.046
	28.00	.417	5.791	7.390	5.666	5⅝	5.625	.041
	32.00	.475	5.675	7.390	5.550	5⅜	5.375	.175
	17.00	.231	6.538	7.656	6.413	6⅜	6.375	.038
	20.00	.272	6.456	7.656	6.331	6¼	6.250	.081

TABLE 3.6 *(continued)*

	Dimensions					Bit Size and Diametral Clearance		
Size OD	Weight Per Foot Nominal	Wall	Nominal Coupling ID	Nominal Coupling OD	Drift	Nominal Bit Size		Clearance From Drift Dia.
Inches	Pounds	Inches				Inches	Decimal	Inches
	23.00	.317	6.366	7.656	6.241	6⅛	6.125	.116
7	26.00	.362	6.276	7.656	6.151	6⅛	6.125	.026
	29.00	.408	6.184	7.656	6.059	6	6.000	.059
	32.00	.453	6.094	7.656	5.969	5⅞	5.875	.094
	35.00	.498	6.004	7.656	5.879	5⅞	5.875	.004
	38.00	.540	5.920	7.656	5.795	5¾	5.750	.045
	20.00	.250	7.125	8.500	7.000	6¾	6.750	.250
	24.00	.300	7.025	8.500	6.900	6¾	6.750	.150
	26.40	.328	6.969	8.500	6.844	6¾	6.750	.094
7⅝	29.70	.375	6.875	8.500	6.750	6¾	6.750	.000
	33.70	.430	6.765	8.500	6.640	6⅝	6.625	.015
	39.00	.500	6.625	8.500	6.500	6⅜	6.375	.125
	24.00	.264	8.097	9.625	7.972	7⅞	7.875	.097
	28.00	.304	8.017	9.625	7.892	7⅞	7.875	.017
	32.00	.352	7.921	9.625	7.796	7¾	7.750	.046
8⅝	36.00	.400	7.825	9.625	7.700	7⅝	7.625	.075
	40.00	.450	7.725	9.625	7.600	7⅜	7.375	.225
	44.00	.500	7.625	9.625	7.500	7⅜	7.375	.125
	49.00	.557	7.511	9.625	7.386	7⅜	7.375	.011
	29.30	.281	9.063	10.635	8.907	8¾	8.750	.157
	32.30	.312	9.001	10.625	8.845	8¾	8.750	.095
	36.00	.352	8.921	10.625	8.765	8¾	8.750	.015
9⅝	40.00	.395	8.835	10.625	8.679	8⅝	8.625	.054
	43.50	.435	8.755	10.625	8.599	8½	8.500	.099
	47.00	.472	8.681	10.625	8.525	8½	8.500	.025
	53.50	.545	8.535	10.625	8.379	8⅜	8.375	.004
	32.75	.279	10.192	11.750	10.036	9⅞	9.875	.161
	40.50	.350	10.050	11.750	9.894	9⅞	9.875	.019
	45.50	.400	9.950	11.750	9.794	9¾	9.750	.044
10¾	51.00	.450	9.850	11.750	9.694	9⅝	9.625	.069
	55.50	.495	9.760	11.750	9.604	9	9.000	.604
	60.70	.545	9.660	11.750	9.504	9	9.000	.504
	65.70	.595	9.560	11.750	9.404	9	9.000	.404
	38.00	.300	11.150	12.750	10.994	10⅝	10.625	.369
	42.00	.333	11.084	12.750	10.928	10⅝	10.625	.303
11¾	47.00	.375	11.000	12.750	10.844	10⅝	10.625	.219
	54.00	.435	10.880	12.750	10.724	10⅝	10.625	.099
	60.00	.489	10.772	12.750	10.616	9⅞	9.875	.741
	48.00	.330	12.715	14.375	12.559	12¼	12.250	.309
	54.50	.380	12.615	14.375	12.459	12¼	12.250	.209
13⅜	61.00	.430	12.515	14.375	12.359	12¼	12.250	.109

TABLE 3.6 *(continued)*

Dimensions						Bit Size and Diametral Clearance		
Size OD	Weight Per Foot Nominal	Wall	Nominal ID	Coupling OD	Drift	Nominal Bit Size		Clearance From Drift Dia.
Inches	Pounds	Inches				Inches	Decimal	Inches
	68.00	.480	12.415	14.375	12.259	12¼	12.250	.009
	72.00	.514	12.347	14.375	12.191	12	12.000	.191
	55.00	.312	15.375	17.000	15.188	15	15.000	.188
16	65.00	.375	15.250	17.000	15.062	15	15.000	.062
	75.00	.438	15.125	17.000	14.938	14¾	14.750	.188
	84.00	.495	15.010	17.000	14.823	14¾	14.750	.073
20	94.00	.438	19.124	21.000	18.936	17½	17.500	1.436

NOTE: Above information for API casing. For other casing data, refer to Toolpushers' Manual or specific manufacturer's specifications. For metric equivalents (millimeters), multiply the "inches" by 25.4. Source: Courtesy of the Security Division of Dresser Industries, Inc.

API STANDARD BIT TOLERANCE

Bit Size—Inches	Tolerance—Inches
3⅜–13¾	+1/32–0
14 –17½	+1/16–0
17⅝ & larger	+3/32–0

TABLE 3.7 CASING JOB FORM

__________CASING JOB

T.D. OF HOLE____________________SIZE HOLE____________________

LAST CSG. SIZE__________________DEPTH SET____________________

MUD WT. IN HOLE_________________EST. FRAC. GRAD.____________

PURPOSE FOR WHICH CSG. IS BEING RUN:_________________________

HOLE PROBLEMS TO CONSIDER:___________________________________

D.C. SIZE AND TYPE BHA USED TO DRILL HOLE:___________________

CSG. DESCRIPTION:____

CSG. INSPECTION:____

SPECIAL DRIFT SIZE IF NEEDED:________________________________

METHOD OF MEASURING CSG. ON RACK AND FOR RUNNING:____________

SPECIAL HANDLING INSTRUCTIONS:_______________________________

FLOAT EQUIP:____

CENTRALIZERS: NO., TYPE, AND SPACING:________________________

TYPE THREAD LUBRICANT:_______________________________________

MISC. EQUIPMENT:____

RUNNING EQUIPMENT:____

Elev. slips:____

Tongs:____

Rams:____

Safety Valve:____

Etc.:____

RUNNING SPEED:____

FILL INSTRUCTIONS:____

CSG. PHYSICAL DATA:____

Make-up Torque:______________________________________

Disp. (bbls/lin.ft)_________ Cap. (bbls/lin.ft)_________

Max. Safe Pull: Body @ 1.6 SF.____Coupling @ 1.6 SF____

Collapse @ 1.0 SF___________________psi

Burst @ 1.0 SF___________________psi

DRIFT I.D.:_______________IN. NOMINAL I.D.:_______________IN.

PRIOR TO RUNNING CSG:

____ Cement blended and tested on rig provides required thickening time and tests confirm lab. tests.

____ Adequate fluid on hand to pump down plug should loss circulation occur.

____ Cementing head loaded.

____ Swage and safety valve on floor.

TABLE 3.7 *(continued)*

____ Extra tongs checked out and ready.

____ Running equipment inspected.

CASING RUNNING PROCEDURE:

____ Thread lock bottom____jts. w/_____thread lock.

____ Place_____jts. between shoe and collar.

____ Place centralizers as instructed.

____ Fill csg. and check circulation through shoe after picking up_____jts.

____ Make up csg. to________ft. lbs. torque.

____ Advise tong operator of different wts./grades that would require different torque. (also x-over jts.).

____ Csg. to leave shoe of last csg. string @ jt. No.___ Check csg. for complete fill and make any adjustments or repairs necessary before going into open hole.

____ Landing joint and subsea running tool prepared and ready to install in drill pipe string.

____ Check string weight against drill pipe running string require minimum of 100,000 lb. over pull or 2.5 s.f., whichever higher.

____ Circ. twice cap. of annulus, 2 hours, or until flow line temperature stabilizes, whichever is greater, before cementing. (revise if losing circulation).
Circ. @ _______________bpm.
Cap. of csg. = _______________bbls.
Cap of ann. = _______________bbls.

____ Est. wt. of csg. in mud_______________#.
Act. wt. of csg. in mud_______________#.

CSG. DETAIL:

Item	Description	Section Length	Cumulative Length
Shoe			
Jts.			
Collar			
Jts.			
Jts.			
Jts.			

TABLE 3.7 *(continued)*

Total csg. run ____________
Running string to surface kb ____________
csg. setting depth at kb ____________

Mud displaced as csg. is run: __________bbls/100':
__________: __________bbls. total
Cement blend: yield: water req:_____________________

PUMPING TIMES:

	LEAD SLURRY	TAIL SLURRY
Desired:	____hrs.____min.	____hrs.____min.
Lab. test:	____hrs.____min.	____hrs.____min.
Prejob test:	____hrs.____min.	____hrs.____min.

COMPRESSIVE TEST: (Tail Slurry)

Desired:	___psi @ 8 hr.	___psi @ 12 hr.	___psi @ 24 hr.
Lab. test:	___psi @ 8 hr.	___psi @ 12 hr.	___psi @ 24 hr.
Prejob test:	___psi @ 8 hr.	___psi @ 12 hr.	___psi @ 24 hr.
Postjob test:	___psi @ 8 hr.	___psi @ 12 hr.	___psi @ 24 hr.

Desired cement top:_____________ft.

CEMENT CALCULATIONS:

Cap.___ft.___"x___" ann. @___bbls/lin. ft. = ________bbls.
Excess for open hole = _____% x _____bbls. = ________bbls.
csg.
Cap.___ft.___ x___" ann. @___bbls/lin. ft. = ________bbls.
Total vol. to be filled w/cement = ________bbls.

SX. NEEDED:

Lead - _____bbls. x 5.6 = ft^3 ÷ _____ft^3/sx = _________sx.
Tail - _____bbls. x 5.6 = ft^3 ÷ _____ft^3/sx = _________sx.
Cement mixing procedure and special instructions:_________
____

Precede cement w/_____bbls._____ppg._____________mud flush
Pump_____bbls. cement. wt._____ppg. @ __________bbls./min.
Drop wiper plug - w/_________or wo/_________cement on top.

TABLE 3.7 *(continued)*

DISP. CEMENT W/:

_______bbls._______ppg._______spacer @ _______bbls./min.
_______bbls._______ppg._______fluid @ _______bbls./min.
_______bbls. total.

DISP. CALC (CAPACITY OF CSG. TO COLLAR):

_______ft._______ #@ _______bbls./lin.ft. = _______bbls.
_______ft._______ #@ _______bbls./lin.ft. = _______bbls.
_______ft._______ #@ _______bbls./lin.ft. = _______bbls.
_______ft._______ #@ _______bbls./lin.ft. = _______bbls.
_______ft. -------------TOTAL------------- _______bbls.

CEMENT TO LEAVE SHOE W/_____BBLS. OF CEMENT OF DISPLACEMENT PUMPED. BUMP PLUG W/_____PSI ABOVE HYDROSTATIC DIFF. RELEASE PRES. AND CHECK FLOAT EQUIP. (IF NOT HOLDING APPLY_____PSI ABOVE HYDROSTATIC DIFF. AND LET SET______ HRS.). MAXIMUM SURFACE PRESSURE SHOULD CEMENT PLUG OR SQUEEZE OFF.___________PSI.

COMMENTS:____

TABLE 3.8 LINER JOB FORM

__________ LINER JOB

LINER WILL BE SET AT____
TOP OF LINER TO BE AT APPROX.____________________________ft.
SHOE OF LAST CASING AT___________________________________ft.
HANGER TO BE SET IN_______", _______#, _________GRADE CASING.
MUD WT.______PPG. EST. FRAC. GRAD. AT LAST CSG. SHOE_____PPG.
COMMENTS:____

CHECK DRILL STRING DESIGN TO RUN LINER. REQUIRE 1,000,000 lb OVERPULL OR 2.5 SAFETY FACTOR, WHICHEVER GREATER.

CHECK RATINGS OF DRILL LINE, BLOCK, HOOK COMPENSATOR, HANDLING EQUIPMENT, ETC.

LINER DESCRIPTION:____

TABLE 3.8 *(continued)*

INSPECTION:______________________________

TYPE HANGER:______________________________

SPECIAL HANDLING INSTRUCTIONS:____________________

MAKE-UP TORQUE:______________________________

FLOAT EQUIPMENT:______________________________

CENTRALIZERS: NO, TYPE, AND SPACING:________________

TYPE PLUG DROPPING HEAD:________________________

TYPE THREAD LUBRICANT:________________________

RUNNING EQUIPMENT:

- elevators:______________________________
- slips:______________________________
- safety clamp:______________________________
- tongs:______________________________
- rams:______________________________
- lift plugs:______________________________
- safety valve:______________________________
- etc.:______________________________

PHYSICAL DATA:

- disp. (bbls./lin.ft.__________cap. (bbls./lin.ft)________
- max. safe pull: body @ 1.6sf.____# joint @ 1.6sf._____#
- collapse @ 1.0 s.f. ____________________psi
- burst @ 1.0 s.f. ____________________psi
- drift i.d.____________in.: nominal i.d.____________in.
- body o.d.____________in.: joint o.d. ____________in.

FILL INSTRUCTIONS:______________________________

RUNNING SPEED:______________________________

LINER DETAIL:

________________________type shoe________________

____________jts.__________liner____________________

TABLE 3.8 *(continued)*

____________________________landing collar______________

_____________jts.___________liner_______________________

____________________________hanger______________________

____________________________setting tool________________

_______stds. of_______d.p.______________________________

_______stds. of_______d.p.______________________________

_______singles of________d.p.___________________________

_______pup jts. of________d.p.__________________________

total d.p. & liner ____________ ____________

less depth of hole ____________ ____________

d.p. above rotary total ____________ ____________

liner weight in mud (with liner full) (less block)___lbs.

LINER TO LEAVE LAST CASING SHOE ON STD. NO. _________________.

TAG BOTTOM (UNLESS RISK IS TOO GREAT). HANG LINER.

hanging procedure:_______________________________________

CIRC. CAP. OF D.P./LINER OR ANNULUS - WHICHEVER IS GREATER

CAP. D.P./LINER - __________BBLS.

____ft. of ____ @ ____ bbls./lin. ft. = ______bbls.

____ft. of ____ @ ____ bbls./lin. ft. = ______bbls.

____ft. of ____ @ ____ bbls./lin. ft. = ______bbls.

____ft. of ____ @ ____ bbls./lin. ft. = ______bbls.

CAP. ANNULUS - __________BBLS.

____ft. of ____"x____"@____bbls./lin. ft. = ____bbls.

____ft. of ____"x____"@____bbls./lin. ft. = ____bbls.

____ft. of ____"x____"@____bbls./lin. ft. = ____bbls.

____ft. of ____"x____"@____bbls./lin. ft. = ____bbls.

RELEASE FROM HANGER - PICK UP_____________FT. TO CHECK RELEASE.

SET________# DRILL STRING WT. ON HANGER FOR CEMENTING.

CEMENT BLEND:____

____

PUMPING TIME TEST	LEAD SLURRY	TAIL SLURRY
desired:	______hrs.______min.:	______hrs.______min.
lab. test:	______hrs.______min.:	______hrs.______min.
prejob test:	______hrs.______min.:	______hrs.______min.
postjob test:	______hrs.______min.:	______hrs.______min.

TABLE 3.8 *(continued)*

COMPRESSIVE TEST	TOP OF LINER________0'	BOTTOM OF LINER________0
desired:	18____24____36____	18____24____36____
lab. test:	18____24____36____	18____24____36____
prejob test:	18____24____36____	18____24____36____
postjob test:	18____24____36____	18____24____36____

MIX________ SX.________BBLS. CEMENT IN BATCH MIXERS.

cap.____ft.____"X____"ann. @ ____bbls./lin.ft. =____bbls.
excess for open hole = ____% X ____bbls. =____bbls.
cap.____ft.____"X____"ann. @ ____bbls./lin.ft. =____bbls.
cap. above liner = ____ft. @ ____bbls./lin.ft. =____bbls.
cap. shoe jts. = ____ft. @ ____bbls./lin.ft. =____bbls.

total volume to be filled w/cement =____bbls.

PRECEDE CEMENT W________BBLS.___________MUD FLUSH

PUMP________BBLS. CEMENT SLURRY @__________BBLS./MIN.

DROP D.P. WIPER PLUG.

DISPLACE WITH ________BBLS.
__________bbls.___________ @ _________bbls./min.
__________bbls.___________
cement to leave shoe w/__________bbls. disp.
dart to pick up liner wiper w/__________bbls. disp.

BUMP PLUG W/________PSI ABOVE HYDRO. DIFF.

MAX. SURFACE PRES. SHOULD CEMENT SQUEEZE OR SCREEN OUT___PSI.

RELEASE PRESSURE AND CHECK FLOAT EQUIPMENT.

HOLD___PSI ON D.P. WHILE PULLING SETTING TOOL OUT OF HANGER.
if d.p. press. falls to zero psi. pph. pooh.
if d.p. press. holds or increases, bleed mud back into howco tank to equalize fluid columns. (pull ______ stands slowly before stopping to slug. d.p.).

<u>PRIOR TO RUNNING LINER</u>

_____ cement blended and tested
_____ correct rams in b.o.p.'s if needed.

TABLE 3.8 *(continued)*

_____ adequate fluid on hand to pump down plug should loss circ. occur.

_____ Drill string & all subs rabbited.

_____ swage and safety valve on location.

_____ extra tongs checked out.

_____ running equipment inspected.

TABLE 3.9 CASING-TALLY SUMMARY SHEET

DATE: ______

FIELD ______ LEASE & WELL NO. ______ TALLY FOR ____" CASING

SUMMARY OF PAGE MEASUREMENTS			
	NO. OF JOINTS	FEET	.00'S
PAGE 1			
PAGE 2			
PAGE 3			
PAGE 4			
PAGE 5			
PAGE 6			
PAGE 7			
PAGE 8			
PAGE 9			
TOTAL			

SUMMARY OF DEPTH CALCULATIONS				
		NO. OF JOINTS	FOOTAGE	
			FEET	.00'S
1	TOTAL CASING ON RACKS			
2	LESS CASING OUT (JTS NOS.			
3	TOTAL (1 – 2)			
4	SHOE LENGTH			
5	FLOAT LENGTH			
6	MISCELLANEOUS EQUIPMENT LENGTH			
7	TOTAL CASING AND EQUIPMENT FROM CEMENT HEAD (3 + 4 + 5 + 6)			
8	LESS WELL DEPTH			
9	"UP" ON LANDING JOINT			

SUMMARY OF STRING AS RUN								
WEIGHT	GRADE	THREAD	MANUFACTURER	CONDITION NEW-USED	LOCATION IN STRING	NO. OF JOINTS	FOOTAGE	INTERVAL
					JT NO. THRU NO.			'– '
					JT NO. THRU NO.			'– '
					JT NO. THRU NO.			'– '
					JT NO. THRU NO.			'– '
					JT NO. THRU NO.			'– '
					JT NO. THRU NO.			'– '
					JT NO. THRU NO.			'– '

4
Mud Logging

Very often the drilling system will include on-site mud logging equipment and personnel and even digital computer facilities. Such have become a standard feature of drilling operations in many parts of the world, especially in deep wells, offshore locations, remote wildcats, and in areas where special and potentially dangerous drilling conditions may be encountered. Most drilling rigs today are provided with various devices that automatically record such parameters as weight of the drillstring, mud pump pressure, rate of penetration, mudpit fluid level, mud flow rate, etc., all as functions of time.

Facilities such as these serve not only to monitor the overall progress of the drilling operation, but also permit the prediction and rapid assessment of potentially dangerous situations. With the data obtained, calculations can be made to insure a safer and more efficient drilling operation.

Hydrocarbon well logging, better known as mud logging, is a service performed at the rig site with a self-contained mobile field laboratory. These mud logging units are equipped to provide a variety of services, depending upon the requirements of the well, but the two basic services are (1) detection of hydrocarbons in the drilling mud and drill cuttings, and (2) monitoring of the drilling mud density. The former is, of course, primarily for the detection of hydrocarbon-bearing strata while the latter is primarily for blowout prevention.

Modern mud logging units are capable of providing a variety of additional tests and services applicable to essentially all areas of the drilling operation and particularly to the functions of the wellsite geologist. These tests and services are associated with measurements and interpretations of drilling mud and well-cuttings data and the monitoring of various aspects of the drilling operation. The relationships of these services, the data sources, and the reported results are illustrated in figure 4.1. Figure 4.2 shows the recommended standard format that is used by most mud logging operators. Figure 4.3 is a daily summary form.

The mud-logging unit is generally powered by the rig's electrical generator and is connected to the rig in three ways: (1) by a hose-and-pump connection to the mud-return flow line; (2) by cables to moving rig components; and (3) by electrical lines to sensors or recording devices.

4.1 Gas Detection

A logging unit usually contains two separate gas detectors, one for mud (continuous) and one for cuttings (batch). Detection of gas in the returning mud is done continuously and recorded automatically. Simple detection normally employs a hot-wire detector, or equivalent type, which senses the presence of combustible gases. An alarm system is provided to alert the operator to a high gas reading.

In addition to gas detection, most units include a gas chromatograph that separates the gas components in the stream and records the quantity of each on a cyclic basis. Equipment can also be included to detect and analyze for H_2S, CO_2, He, etc. Hydrogen sulfide detectors normally also include an alarm system.

Detection of gas in the cuttings is made at fixed intervals, usually every 5 or 10 ft of drilling, by agitating fresh cuttings and

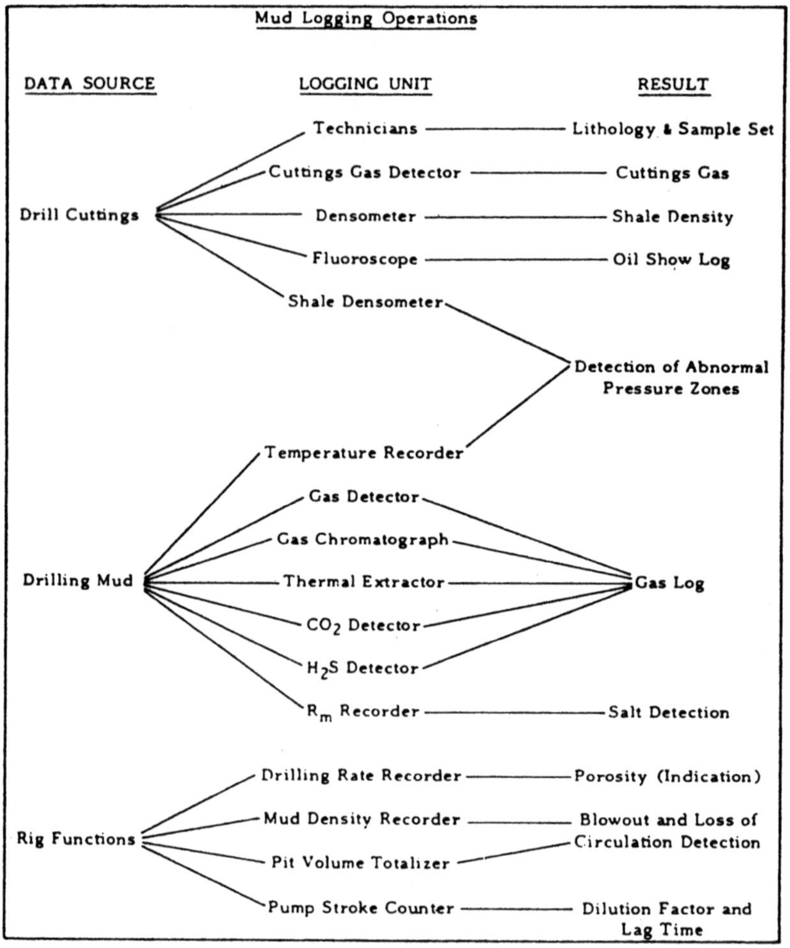

Figure 4.1 Mud Logging Operations.

water in a high-speed mixer. A hot-wire type detector and panel meter are used to indicate the presence of hydrocarbon gases. This instrument determines the relative amount and kind of gas in the drill cuttings and gives an instrumentally obtained evaluation of the oil in the cuttings, both readings being qualitative in nature, rather than quantitative.

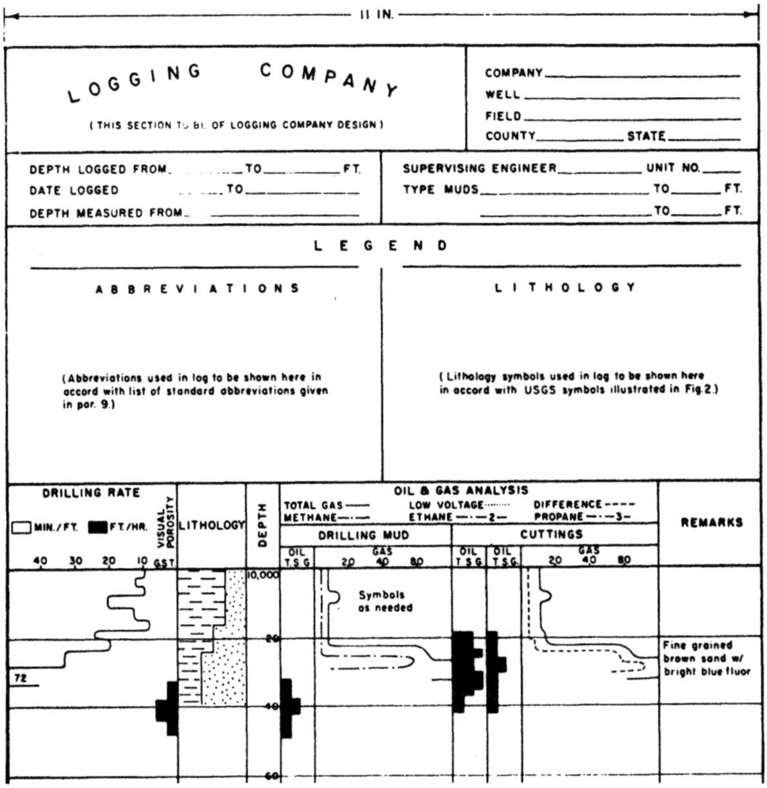

NOTE: The letters "T.S.G." shown in column headings for visual porosity, drilling mud—oil, and cuttings—oil indicate trace, show, or good fluorescence. However, any appropriate scale is acceptable.

Figure 4.2 Recommended standard hydrocarbon mud log form. All curves and legends drawn for illustrative purposes only. (Courtesy of the American Petroleum Institute from *Recommended Standard Hydrocarbon Mud Log Form*, API RP 34: Recommended practice standard hydrocarbon mud log form, first edition, November 1958, American Petroleum Institute, Washington, D.C.)

4.2 Drilling Rate

The drilling rate is obtained from one of two possible sources, both being dependent on measuring the downward movement of the drillpipe as drilling progresses. With a drilling-rate recorder,

MUDLOGGING SERVICE

WELL ______________________ DATE ______________________

DEPTH AT 7:00 A.M. ________________ TRAILER NO. __________

NO. OF FT. DRILLED IN LAST 24 HRS. ______________________

MW _____ VIS _____ WL _____ CL _____ CA _____ PH _____ FC _____

REMARKS:

PRESENT ACTIVITY

PRESENT FORMATION LITHOLOGY %

BACKGROUND GAS

HI ____________ UNITS AT ______________________

LOW ____________ UNITS AT ______________________

AVG. ____________ UNITS AVG. CONNECTION GAS ____________

RATE OF PENETRATION

HI ____________ FT. PER HR. AT ______________________

LOW ____________ FT. PER HR. AT ______________________

AVG. ____________ FT. PER HR.

FRM TOPS:

NEXT FORMATION EXPECTED AT

SHOWS

DEPTH % FLUORESCENCE COLOR CUT FRACTURES

Figure 4.3 Dailly summary form.

such as a Geolograph, a sensor can be installed in the recorder case to transmit a signal to the logging unit as each foot or meter is drilled.

Alternatively, a wireline retriever and measuring wheel can be incorporated in the logging unit. This wireline can be run to the top of the drilling mast and then down to the kelly, so that vertical pipe movement is reflected in a rotation of the measuring wheel.

In either case, with each foot drilled, a mark is made on a continuous chart recorder, a panel depth meter is advanced, and an audible signal alerts the logger that another foot has been drilled. With the time for each foot recorded, the drilling rate, in minutes per foot, can be obtained.

4.3 Pump Stroke Counter

Switches installed on the rig mud pumps close once with each stroke of the pump, thereby operating two independent counters and a pump-rate meter. One counter can be "lagged" behind the other by the number of pump strokes theoretically required to bring drill cuttings to the surface. When a foot has been drilled, the logger makes a note of the counter reading (cuttings on bottom), then when the cuttings-up counter reaches the same value, a sample of cuttings from that foot can be obtained at the shale shaker. In this manner, the lag time of the cuttings is approximated regardless of the speed of the mud pump or even if the pump has to be stopped for a brief period to make a drillpipe connection.

The main purpose of the pump stroke counter is to correct for the lag time of drill cuttings, but to do this correctly, it must be used properly. The "lag" strokes for every major change in the hydraulic system must be calculated; that is, when casing is landed, when hole size is reduced, when pump-liner size is changed, and when changing from one pump to two. Routinely, the lag-stroke increment for each 100 feet of additional hole drilled is calculated and added to the counter setting.

Carbide, rice, or lentil checks can be used to determine the exact number of pump strokes required to bring drill cuttings to the surface in the case of irregularities, such as caverns or an enlarged hole, which will alter the actual lag time from the calculated value.

4.4 Electric Logging

Since we are discussing logging, we will touch briefly on electric logging. The major logging is done while the drilling rig is still over the hole.

Prior to logging, the hole should be in good condition and the mud in excellent shape or no logs or very poor logs will be the end result. A list of logs that can be run and their possible use are presented in table 4.1.

Mud logging, electric logging, and core analysis go hand in hand, so let's touch briefly on coring.

4.5 Coring and Core Analysis

The *American Heritage Dictionary of the English Language* defines a *core* as "the most important part of anything," and *coring* as "to remove the core of." Although it is entirely possible that the editors of this reference work never heard of an oilfield core or of coring as a part of oil and gas well drilling and testing, these definitions are close to the truth.

In the petroleum industry, a core is often defined as any sample of subsurface rock strata obtained from a borehole and of a size large enough to permit the measurement of its important physical properties. In this regard, a core offers the only direct means by which a reservoir rock can be described in terms of its physical characteristics, its fluid content, and its fluid-flow properties. Therefore, it might well be the most important part of the process of reservoir description.

Many physical, chemical, and thermodynamic properties can be defined by core analysis, but the most important information

TABLE 4.1 LOGS AND THEIR POSSIBLE USES

Type Log	Use
Induction electrolog (IEL)	1. Determine formation resistivities 2. Formation evaluation 3. Depth contact 4. Correlation
Dual-induction focused log (DIFL)	1. Formation resistivity in medium- to low-porosity zones 2. Determine depth of invasion 3. Correlation
Dual laterlog (DLL) Laterlog (LL)	1. Formation resistivity in salt drilling fluid 2. Depth control 3. Correlation 4. Formation evaluation
Microlaterlog (MLL)	1. Determination of net pay thickness 2. Location of porous and permeable zones 3. Detection of movable oil 4. Determination of hole size
Proximity minilog (PML)	1. Measurement of flushed zone resistivity (Rxo) in case of thick mudcakes and deeper invasion 2. Determination of hole size and mudcake thickness 3. Detection of movable hydrocarbons 4. Determination of net pay
Borehole-compensated acoustilog (BHC)	1. Porosity in liquid-filled boreholes 2. Accurate porosity in rugose boreholes 3. Correlation 4. Formation velocity data 5. Identify lithology when used with other porosity devices 6. Secondary porosity when used with other porosity devices
Compensated neutron log (CNLOG)	1. Porosity determination 2. Located gas (when combined with CDL)

TABLE 4.1 *(continued)*

Type Log	Use
	3. Identify lithology when used with other porosity devices
Sidewall Epithermal Neutron (SWN)	1. Porosity determination 2. Locate gas–oil contact 3. Identify lithology when used with other porosity devices
Gamma-ray log	1. To define lithology 2. For correlation 3. To indicate shale content 4. Locate radioactive tracers
Neutron log	1. Evaluate porosity 2. Determine lithology 3. Locate gas-bearing formations 4. Correlation and depth control
Acoustic cement-bond log (ACBL)	1. Measure cement effectiveness 2. Evaluate squeeze operations 3. Locate cement top
Spontaneous potential curve	1. To define lithology 2. Locate boundaries between beds 3. For correlation 4. Obtain good values for formation water resistivity

normally desired from a core are porosity, permeability, and fluid saturations, including water, oil, and gas. These three properties can also be obtained by indirect means, such as wireline logs and various types of well tests, but these data are always subject to question without core analysis data for calibration and verification.

Cores are also obtained for a variety of other reasons, including lithological and other types of geological studies, studies of rock fracture patterns, and studies to define or improve well-completion practices. To the geologist, a core offers the only means to see the rock as it appears in its natural state.

4.6 Coring Methods

Through the years, many types of equipment and many different techniques have been devised for the taking of cores, some of which were successful and have evolved into the equipment and techniques of today. Generally, current methods of coring can be grouped into conventional, wireline, diamond, and sidewall coring.

4.6.1 Conventional Coring

The coring assembly for conventional rotary coring consists of a coring bit and a core barrel, both located on the end of the drillstem. The coring bit is designed essentially the same as are standard rotary rock bits, except that it has citting surfaces only on the perimeter. The core barrel is designed to receive and retain the core as the bit drills into the rock. Typically, the core barrel consists of an inner receiver barrel, an outer barrel, a core retainer or catcher, and a pressure-relief valve to vent core barrel pressure to the outside of the drillstem.

Drilling fluid circulates between the inner and outer barrels in order not to flush the core, thereby increasing recovery of the cored rock. For the same reason, some core barrels have a free-floating inner barrel, which is free to rotate or remain still. For use in very soft, friable or unconsolidated rock, the inner barrel may consist of a heavy rubber tube, or sleeve, which is removed at the surface with the cored formation intact on the inside.

The chief advantages of conventional coring is that a large-diameter core (of 5 or more inches) can be obtained and that, with proper techniques, generally good recoveries can be expected. The chief disadvantage of this method is that the drillstem must be removed to attach the core barrel and bit and then removed again to recover the core and resume normal drilling.

4.6.2 Wireline Coring

This type of equipment permits intermittant coring and drilling and wireline retrieval of cores, all without tripping of the drillstem. All that is required is the installation of a core bit on the end of the drillstem. When a core is to be taken, a retrievable core barrel can be pumped down the drill pipe to a locking seat at the bit. After the core has been cut, the core barrel assembly with the core can be retrieved with a wire-line socket.

When straight drilling is required, a drilling centerbit assembly can be pumped down to seat and lock in the core bit, thereby providing a full cutting head. This can be removed with the wireline when another core is desired.

The major disadvantages of this method are that only a relatively short small-diameter core can be obtained, and core recovery is not as good, on the average, as with more conventional methods. The chief advantage is that a hole can be drilled or a core can be taken as desired and without tripping the drillstem.

4.6.3 Diamond Coring

This method is used to increase core recovery and coring rate, primarily in hard, dense formations. It is generally the same as conventional coring, except that the coring bit is faced with a hard-metal matrix in which are imbedded a large number of industrial-grade diamonds.

The chief advantages in the use of diamond bits are the faster rate of penetration and a longer bit life, permitting up to 90 ft or more of core to be taken before removing the drillstem. Even though these bits may cost 15 to 20 times as much as a conventional bit, they can prove to be economically attractive in many areas.

4.6.4 Sidewall Coring

This is a supplementary method used to obtain core samples from previously drilled zones. The tool is designed somewhat like a perforation tool, but instead of a bullet, a short tube is driven into the side of the borehole and retrieved by means of a cable attached to the body of the tool.

One advantage of this method is obvious: samples can be

obtained from zones which, for one reason or another, were not cored at the time of initial penetration. Also, with the aid of wireline logs, these samples can be taken at very precise positions in the hole.

Disadvantages of this method are numerous. The samples are quite small (approximately ¾ to 1¼ inches in diameter to 2½ inches in length), recovery is often poor, samples are usually broken into smaller pieces either by the capturing tool or by the previous drilling operation, and the rock is taken from that portion of the hole most invaded and disturbed by the drilling fluid. Although they offer a means of sampling a noncored formation, it is seldom that sidewall samples are as satisfactory as a full-hole conventional core. Still, in most cases they are obviously better than nothing.

4.6.5 Reverse Circulation

This method was developed as a means of reducing the number of trips necessary in conventional coring or the use of the wireline for core barrel retrieval in wireline coring. In this method, the mud is circulated down the annulus and then back up the drillstring. Theoretically at least, all of the penetrated formation, including the drill cuttings, will enter the pipe and travel upward to the surface. Usually, a core barrel is kept in the drillstring at the surface where it can be removed every five or ten feet to recover the collected core and cuttings.

Although it works quite well in some instances, this method has not gained wide acceptance.

4.7 Pressure Coring

Pressure coring is a proven method of determining porosity, permeability, water saturations and hydrocarbon saturations. The engineer/geologist should have a core that represents the formation as closely as possible. Pressure coring and a recent development of the sponge barrel helps do this.

The pressure core barrel has the outer barrel with the inner barrel attached to it by means of a swivel assembly which holds the inner barrel in a stationary position as it receives the core. The slip-joint release is activated, when the core is cut, disconnecting the outer barrel which moves downward to seal the upper portion of the pressure-retaining section. This simultaneously activates the ball valve to a closed position at the lower end, thereby sealing off the core at the formation pressure.

Care must be taken during the retrieving operation to be certain that once the barrel has started upward, it is not lowered at any time. It should not be rotated while coming up the hole, as these actions could cause "O" ring damage that might result in a loss of pressure in the core chamber.

When the barrel is out of the hole, the upper section is removed and the captured pressure in the core chamber is checked. The lower assembly is packed in dry ice for approximately 6 hours to immobilize the core saturations. The barrel can be depressurized to atmospheric pressure and the frozen inner core barrel extracted. The inner barrel and core are cut into sections and labeled.

Two problems can effect coring and prevent it from giving much more accurate saturation data:

1. Mud-fluid filtrate invasion of cores; this influx of fluid washes the pore space and can remove a large fraction of the original oil in the cores.
2. The gas saturation of the cores expand as the mud pressure is released when the core is brought to the surface; this expansion causes the core to "bleed", removing a large fraction of the oil from the core.

Sponge cores, like pressure cores, eliminate only one of the two problems of coring, and that is bleeding. The sponge core barrel, except for the inner barrel, is a conventional core barrel. The inner barrel is 30 ft long and developed to accept the 5-ft standard sponge sections. The section is machined on each end. One end has a male fitting and the other a female fitting. This permits the 5-ft sections of sponge to be glued together to make one complete 30-ft section when placed in the inner barrel.

The 5-ft sponge section is drilled through with $\frac{1}{16}''$ holes to allow the gas pressure from the core to escape; the sponge section has proven to be successful in most cases in capturing oil which bleeds from the cores.

The sponge absorbs mud filtrate from the mud as the core barrel goes into the hole. The sponge is preferentially oil-wet, that is, it prefers to be wet with oil rather than being wet with water. The core is forced into the polyurethane sheath during the coring operation. Generally, bottomhole pressure (mud column weight) is higher than the formation pressure. This retains the oil in the core. When the coring operation is complete and as the core barrel is removed from the hole, the pressure on the core is reduced. The dissolved gas in the core expands, causing the oil to "bleed" from the core. Since the sponge is preferentially oil-wet, it rejects the mud filtrate and retains the oil for later analysis.

Once on the surface and placed in tight-fitting PVC containers, the core in the polyurethane sheath is cut into 5-ft sections. Before sealing, formation fluid (salt water) is used to completely fill the containers to preserve core conditions.

The PVC and polyurethane sponge are slit at the laboratory by using a circular saw. The cores are analyzed in the normal manner. The corresponding sponges are analyzed by vaccum distillation at 450 °F. The oil saturation is the sum of the oil in the core plus the oil in the corresponding sponge.

Many of the world's oil fields are candidates for tertiary recovery, and in many cases, good oil-saturation data is not available. When care is taken to use proper coring mud, the best possible oil-saturation data can be obtained. Table 4.2 is a coring check list to assist the engineer and coring personnel in preparing for coring. Figure 4.4 is designed to assist the driller in using the proper weight on the core bit. Figure 4.5 is used to assist the driller in using the proper rotary speed for various formations. Figure 4.6 is an indication of the flow rates with bit size. Flow rates will vary with mud weight and will increase with lighter mud. Most curves are developed using 12.0 ppg drilling fluid.

TABLE 4.2 CORING CHECKLIST

Well Character	Rock Character
Exploratory	Lithology
New-basin strat test	Sandstone
Rank wildcat	Conglomerate
Field extension	Shale
Development	Carbonate
Infill	Chert
Close offset	Evaporite
Stepout	Volcanic
New zone	
Reservoir data	**Physical Character**
Residual saturations	
Fluid contact	Consolidation
Complex Sedimentation	Strength
Location	Abrasiveness
Land	Sticky
Offshore	Brittle
Remote	Fractured
	Vuggy

Gage Hole

Maintain drilled gagehole:

a) Use large drillcollars in relation to roller bit size.
b) Use well stabilized bottomhole assemblies—packed-hole, hookup where possible.
c) Minimize hole doglegs.

Avoid reaming with diamond bit:

a) Select core bit size that will not require reaming.
b) Reduce core bit size if gage trouble expected.
c) Consider reaming run with roller bit prior to coring.
d) Diamond core bit manufacturing to clearance is a minus; if hole is gaged with roller bit it should clear core bit okay.
e) Use full-hole-size core bit where possible to provide better core barrel stabilization and to reduce reaming behind the core bit run.
f) Consider the use of a shock-sub to increase bit life.

Clean Hole

Maintain clean hole for coring:

a) If junk is lost while drilling, fish out immediately to prevent embedding in hole wall.
b) Keep track of tong dies and other miscellaneous iron.
c) Use pipe wiper while tripping for bits or with core barrer.
d) Run junk basket to clean hole on last two roller bit runs.
e) If in doubt, run magnetic tools as required.
f) Wash core bit to bottom *without rotating* to pump remaining junk or rock off bottom before coring begins.

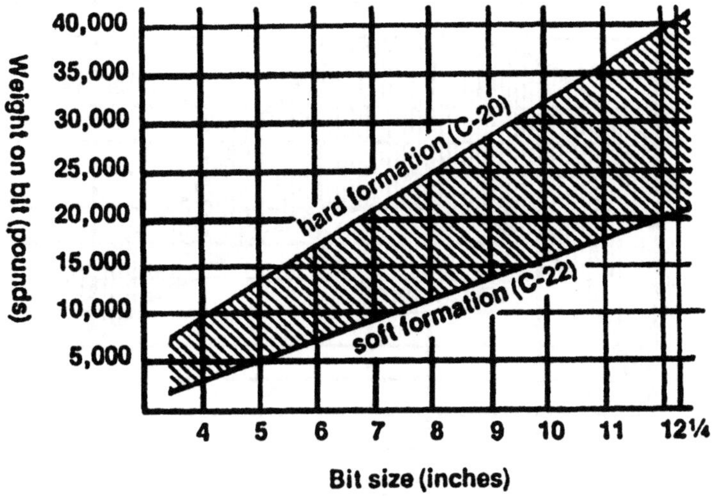

wt. on bit = Number of face stones x K

Diamond Quailty	K Max.	K Min.
SP	30	10
P.Q.	25	10
Select	18	10
Standard	15	10

Figure 4.4 Recommended weight on core bits.

4.8 Hydraulics

Selection of jet nozzles for rock bits should be made to provide maximum expenditure of available energy at the bit.

The annular pressure losses are based upon turbulent flow. When working a hydraulics program, pressure-loss values are corrected for mud weight and not viscosity, since viscosity has

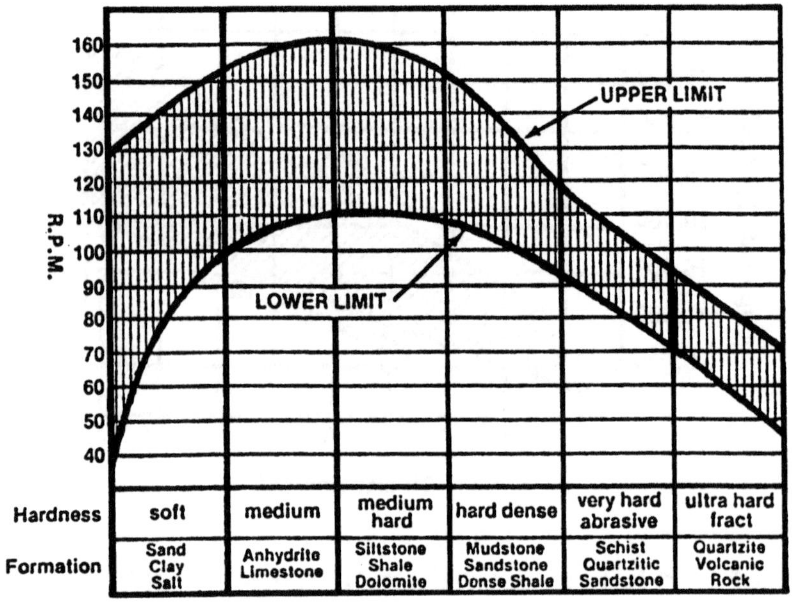

Figure 4.5 Recommended rotating speed for core bits.

very little effect on pressure losses in a turbulent flow. A 200% increase in viscosity will make only a 10% increase in pressure loss, while a 10% increase in mud weight will make a 10% increase in pressure required. Figure 4.7 illustrates a simplified rig circulating system.

Table 4.3 shows a completed Dresser Industries hydraulic worksheet. The tables used to fill out the worksheet can be found in Appendix A at the back of this book.

We now refer back to the Drilling Rules of Thumb (table 1.15):

Flow rate (circulation rate) = 30 to 50 gpm/in. of bit diameter.

Maximum bit hydraulics = 50 to 65% of available pump pressure across the jet nozzles.

Optimum penetration = 4 to 4½ hp/sq. in. of bit size at nozzle velocity of 350 to 400 ft/sec.

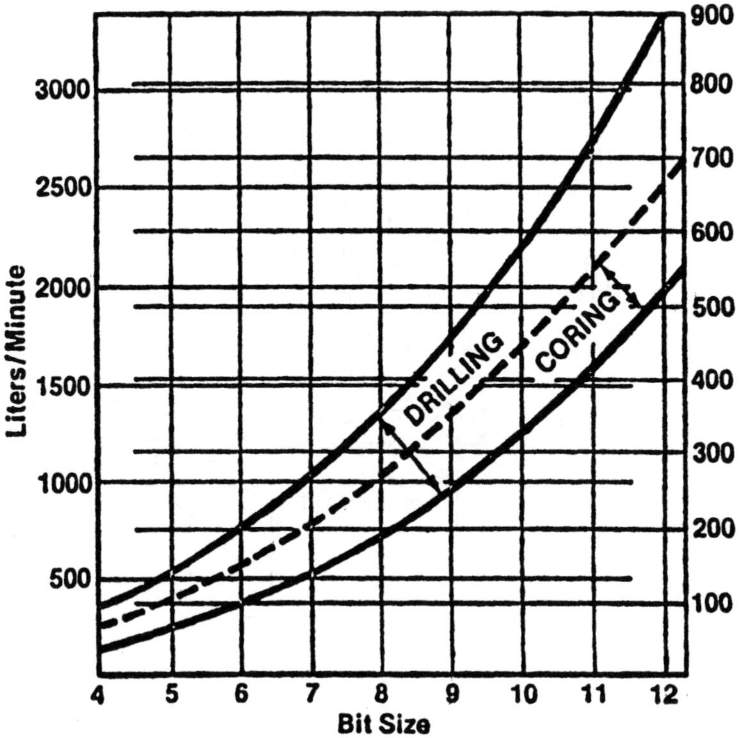

Figure 4.6 Recommended flow rates. Actual flow rate should be adjusted according to mud weight and consistent with good core recovery with regard to formation characteristics.

4.8.1 Basis of Hydraulics Program

The mud pumps are the source of hydraulic energy for the drilling operation supplying the necessary pressure and liquid volume for cleaning the bottom of the hole and removing the rock cuttings from the wellbore. By pumping fluids with a large percentage of solids that have various ranges of viscosity and density over a wide spectrum of pressures and flow rates, the pumps perform under severe operating conditions. For this reason, reciprocating piston pumps should be used because of their rugged construction and simplicity of maintenance. Two types are currently

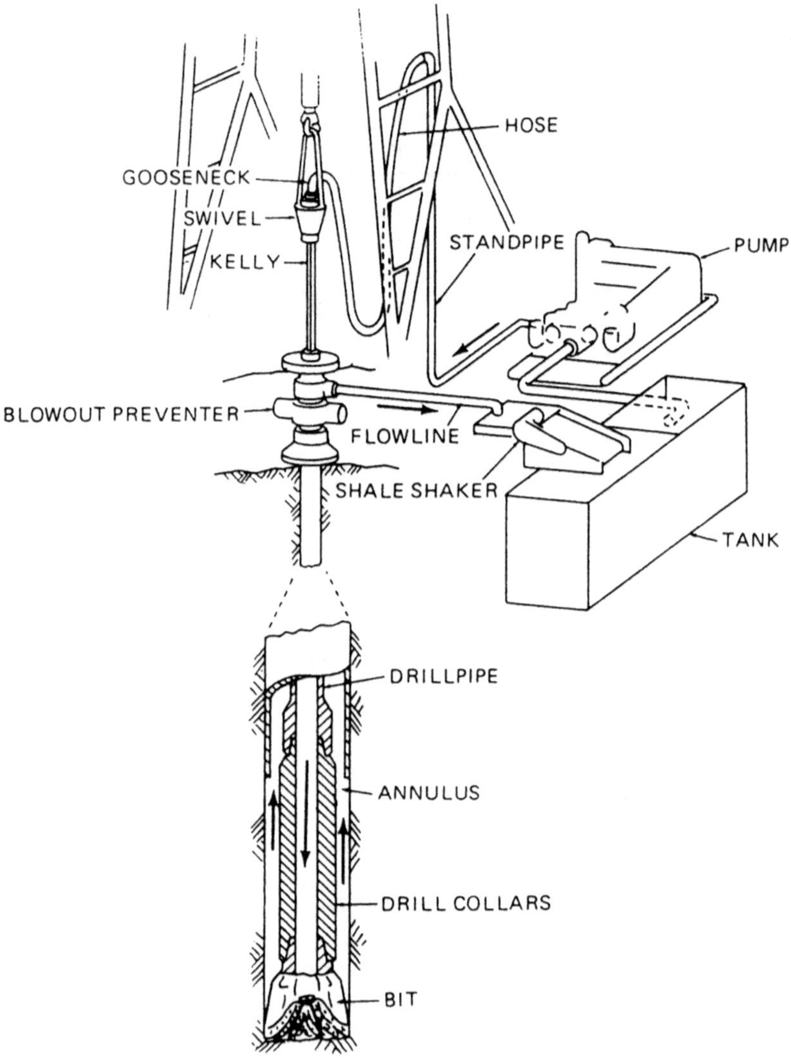

Figure 4.7 Rig circulation system. (Courtesy of PETEX, The University of Texas at Austin)

TABLE 4.3 HYDRAULICS WORKSHEET

Security Operations □ Oilfield Products Division □ Dresser Industries, Inc. □ Post Office Box 6504 □ Houston, Texas 77005 (713) 784-6011.

OPERATOR J. Frantz Petr. Co. CONTRACTOR ACE Drilling RIG NO. 15 DATE 5/15/75

WELL NAME ST. LSE. 11543 NO. D-43 COUNTY Brazos STATE Miss.

WELL DATA:

Hole size 8½" from 7500 to 8500 Drill pipe size 4½" wt. 16.60 t.j. type XH length 8000 ft.

Drill collar length 500 ft. OD x ID 6¾" x 2¼" Mud wt. 9.5 LB/GAL Min. annular velocity (drill pipe) 120

PUMP DATA:

Make IDECO	Liner size 6"	/	/	/
Model MM-1000 GB	Press. Rating 3280	/	/	/
SPM 49	Oper. Press. limit 2500	/	/	/

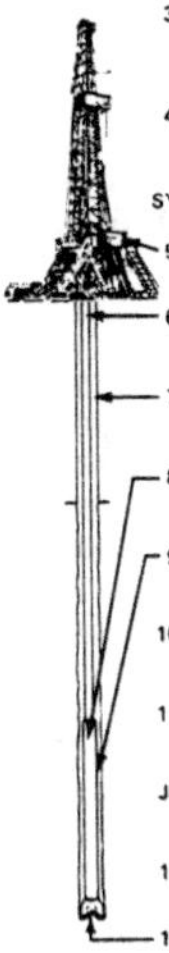

1. Maximum Operating Pressure (Table 1)				2500	psi
Volumetric Discharge (Table 1)				6.8	gal/stk
2. Circulation Rate (Table 2A or 2B)				300	gpm
3. Annular Velocity: (a) Drill pipe (Table 3A)				141	ft/min
(b) Drill collars (Table 3B)				276	ft/min
4. Surface Equipment Type (Table 4)				3	
SYSTEM PRESSURE LOSSES:					
5. Surface Equipment (Table 5)				17	psi
6. Drill Pipe Bore (Table 6): loss per 1000 ft. X length	38/1000	X	8000	= 304	psi
7. Drill Pipe Annulus (Table 7): loss per 1000 ft. X length	8/1000	X	8000	= 64	psi
8. Drill Collar Bore (Table 8): loss per 100 ft. X length	47/100	X	500	= 235	psi
9. Drill Collar Annulus (Table 9): loss per 100 ft. X length	6/100	X	500	= 30	psi
10. System pressure loss (excluding nozzles): add lines 5 thru 9 X Mud Wt./10	650	X	9.5/10	= 618	psi
11. Pressure available for nozzle selection: line 1 minus line 10 X 10/Mud Wt.	1882	X	10/9.5	= 1981	psi
JET NOZZLE SELECTION:					
12. Jet Nozzle Size (Table 10)				9-10-10	32nd
13. Pressure loss through jet nozzles (Table 10): pressure loss X Mud Wt./10	1784	X	9.5/10	= 1695	psi
14. Jet Velocity (Table 11)				445	ft/sec
15. Total pressure expenditure for system: (add line 10 and line 13)				2313	psi
16. % HHp at bit: line 13/line 15 X 100:	1695/2313		X 100	= 73	%

Source: Security Operations, Oilfield Products Division, Dresser Industries, Inc. Reprinted with permission.

being used: duplex (two cylinder) double-acting and triplex (three cylinder) single-acting pumps. PZ pumps are of the latter design.

The pressure and volume range of a given pump is controlled by the particular combination of piston-liner size and the pumping speed (strokes per minute, SPM). The volumetric displacement for a single-acting cylinder is given by

$$q = \frac{\pi D^2}{4} \times S$$

where: q = area of piston × piston travel
D = liner size (in.)
S = stroke (in.)

The total output per cylinder as the pump is driven at N revolutions per minute is:

$$Q = \frac{\pi D^2}{4} \times S \times N \times \frac{1}{231} \qquad \frac{\text{(gal)}}{(\text{in}^3)} \tag{4.1}$$

$$Q = 3.4 \times 10^{-3}\ (D^2 \times S \times N)\ \text{gal/min/piston}$$

In the case of a triplex pump, the total flow will be three times the volumetric displacement of one cylinder. In the case of a duplex double-acting pump, a correction has to be included to take into account the rod volume on one side of the piston, yielding the following relation:

$$Q = (2D^2 - d^2)\ S \times N \times 3.4 \times 10^{-3} \tag{4.2}$$

where D = the rod diameter in inches.

Equations (4.1) and (4.2) show the direct relationship between flow rate and pumping speed for a given liner size. However they are ideal cases since the assumption is made that the volume of fluid displaced is equal to the volumetric piston displacement. This is true only if there is no leakage at the valves or piston. The suction system allows complete fillup of the cylinder during the intake cycle, and there is no gas in the liquid being pumped. This case can be approximated, with good maintenance practices, by supercharging the pump manifold with a centrifugal pump and by proper treatment of the fluid upstream from the intake. In general, however, a displacement efficiency in the order of 85 to 95 percent has to be applied to obtain the actual pump displacement. This number is determined periodically since it is necessary for calculation of the flow of fluids being pumped into the well as a function of the number of pump strokes per unit time. It has been observed that generally efficiency increases with decreases of pumping speed, as shown in figure 4.8.

Assuming that the pump intake is at atmospheric pressure P, and that the discharge pressure is P_2, then the mechanical work performed by the pump, per stroke is given by:

$$\text{WORK} = P_2 \times A \times S = P_2 \frac{\pi D^2}{4} \times S.$$

The power required (work/unit time) is given by

$$\text{POWER} = P_2 \times \frac{\pi D^2}{4} \times \frac{S}{T}$$

where T is the time per stroke (min/stroke).

When pumping at a given speed N (strokes per minute) the power required (work per unit time) is given by:

$$\text{POWER} = P_2 \times \frac{\pi D^2}{4} \times S \times N.$$

Where T is the time per stroke, $T = 1/N$ recalling equation (4.1) for flow rate, the power required can be expressed as:

$$\text{POWER} = P_2 \times q$$

or in field units (HP, psi and gpm)

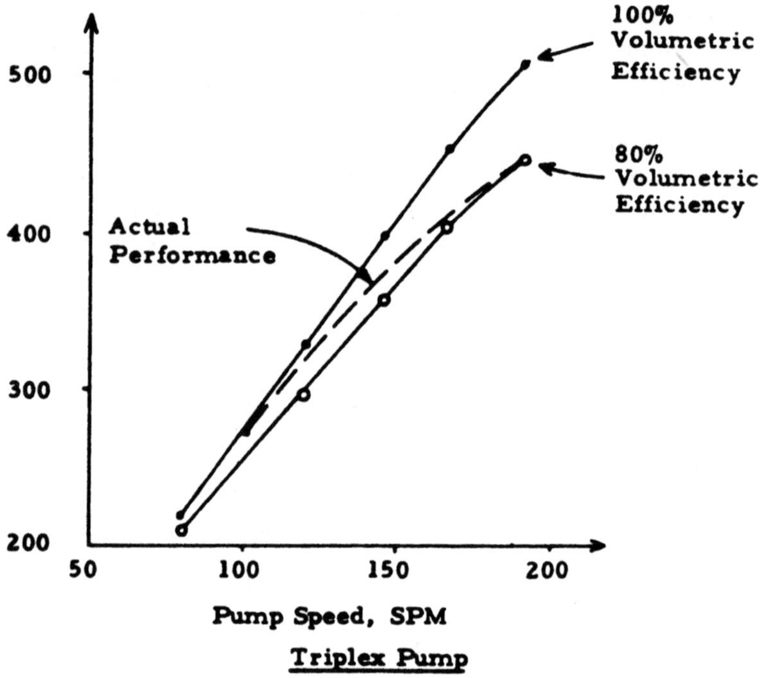

Figure 4.8 Hydraulic horsepower as a function of speed and efficiency.

$$HHP = \frac{P \times Q}{1714}. \qquad (4.3)$$

In general the system is such that it is operated at a given power input level corresponding to the particular speed of the prime mover driving the pump. In practice, therefore, the pump performance relationship (pressure-volume) is expressed by:

$$P \times Q = \text{Constant.}$$

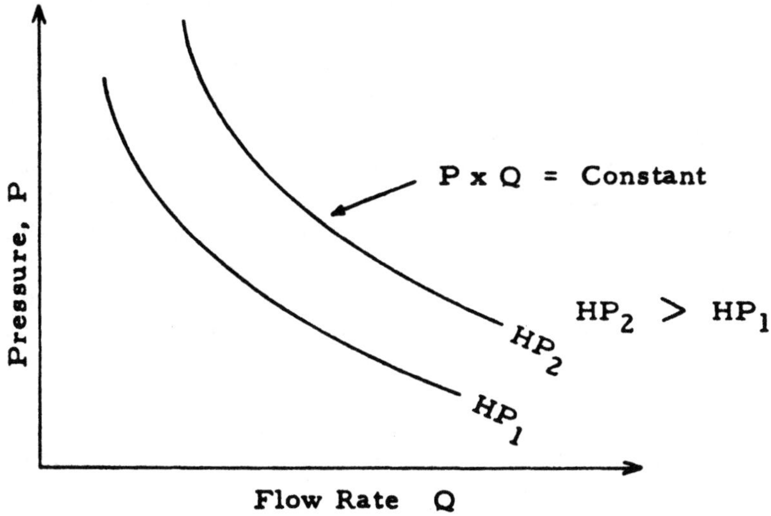

Figure 4.9 Plot of pressure (*P*) vs. flow rate (*Q*).

This represents a hyperbolic relationship indicating that as the flow rate *increases*, the maximum available pressure at that flow rate must decrease as shown in figure 4.9.

For varying levels of input power, families of curves can be generated indicating that as power increases, the available pressure at a given flow rate increases.

Discharge pressure is controlled by plunger diameter, and flow rate is controlled by the combination of plunger diameter and pumping speed. A given pump model is rated at a given *speed and power input* defining the optimum operating conditions for the particular unit, where the overall efficiency and performance is best. For example a Pz-8 triplex (6¼ × 8) is rated at 750 HP at 165 SPM. The stroke is fixed at 8 inches and the maximum piston size is 6¼ inches, for which the volumetric displacement is 528 gallons/minute and maximum discharge pressure is 2200 psi.

$$P = (750)\ (1714)/528 = 2435 \text{ psi.}$$

When applying a power transmission efficiency factor of 90

percent between the prime mover and the pump, the result is the listed discharge pressure of 2200 psi.

Decreasing the plunger size (liner size) to the minimum recommended size of 4 inches results in a reduction flow rate to 215 gallons per minute and an increase of pressure to 5381 psi. The intermediate operating points corresponding to the other liner sizes are represented by the solid line in figure 4.10 for National 7-P-50 triplex. Notice that by changing the operating speed the power requirements change corresponding to the power-RPM relation of the prime mover. All the possible combinations of pumping speed and linear sizes determine the operating range of the pump as indicated by the shaded area in figure 4.11.

Given these operating characteristics, it is possible to operate in two distinct modes depending on the requirements of the drilling operation:

> *Constant input power*, which requires that as the volumetric requirements change, the pressure is varied to maintain the power-available constant. This implies maintaining the speed constant and adjusting the liner sizes.
>
> *Constant-discharge pressure*, which requires that as the volumetric requirements change, the discharge pressure is kept constant by varying the pumping speed for a fixed liner size.

During the drilling history of a well, the volumetric requirements of the circulation system will vary over a wide range. When drilling the surface section, which is generally a large diameter hole, large flow rates will be needed to remove the large volume of cuttings. This requires operating as shown in the right-hand region of figure 4.11 and the mode of operation is at maximum power input.

As depth increases, the shift is towards increased pressure requirements due to the increased pressure losses in the lengthening drill pipe, resulting in constant pressure operation at the maximum level allowable. This maximum is not necessarily determined by the pump performance characteristics (maximum

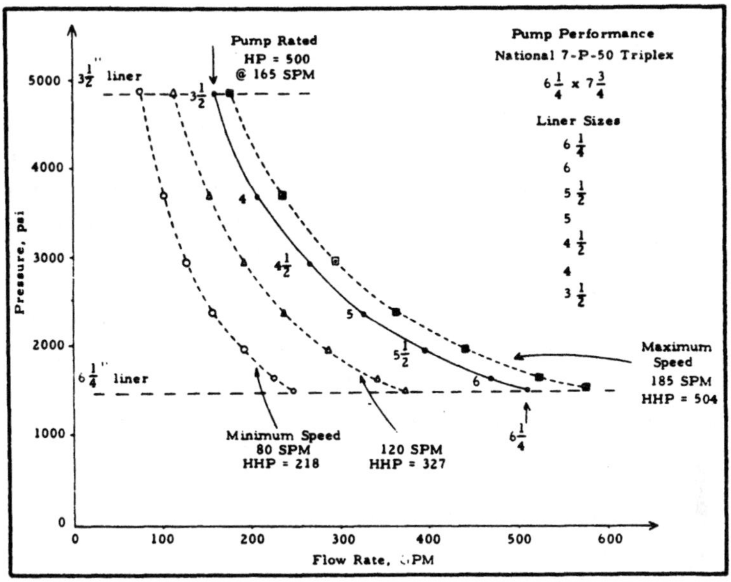

Figure 4.10 Plot of pressure (psi) vs. flow rate (GPM).

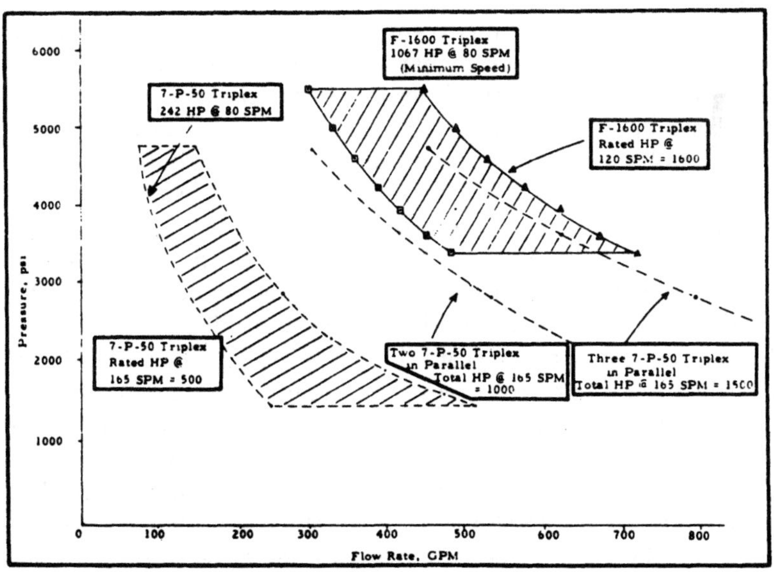

Figure 4.11 Plot of pressure (psi) vs. flow rate (GPM).

discharge pressure) but may be set to a lower value either by the pressure rating of the elements downstream from the pump (standpipe, hose, swivel, etc.), or by agreement with the drilling contractor. (This is common practice in times of unavailability of equipment, since operating and maintenance expenses increase with operating pressure.)

In order to select an adequate pumping system, it is necessary to analyze accurately the hydraulic requirements for the well over the total period of the drilling and completion operations. The well should be "drilled on paper," considering all the possible alternatives and establishing the hydraulic requirements as function of depth. In this manner, it will be possible to closely approximate the real conditions and to estimate the size and power of the pumps required.

Various alternatives can be established that provide more or less advantageous characteristics. For example, the question often arises whether one large pump is better than several smaller pumps connected in parallel. Figure 4.11 illustrates the possibilities and shows that three 500 HP units are almost equivalent to a single 1600 HP pump except for a 10 percent lower discharge pressure and 50 percent higher flow rate at a low pressure. The three units offer considerably greater flexibility as far as pressure-flow rate operating points. Moreover, after the surface hole is drilled and the volumetric requirements decrease, only one or two of the small pumps could be used, resulting in savings in fuel consumption and wear of the equipment. A standby unit would always be available in case of malfunction. In the final analysis, the choice can only be made if the operating requirements can be predicted accurately during the life of the drilling project.

During the last few years there has been a general tendency to shift from duplex to triplex type pumps. Reasons for this are varied, but in general it can be said that this corresponds to the need of higher pressures as well depth increases and a decrease in volumetric requirements. Figure 4.12 shows a comparison between operating characteristics of equal power pumps of the two types. In the case of the triplex pump, increased flexibility and discharge pressures have been obtained at the expense of a lower flow rate at low pressure.

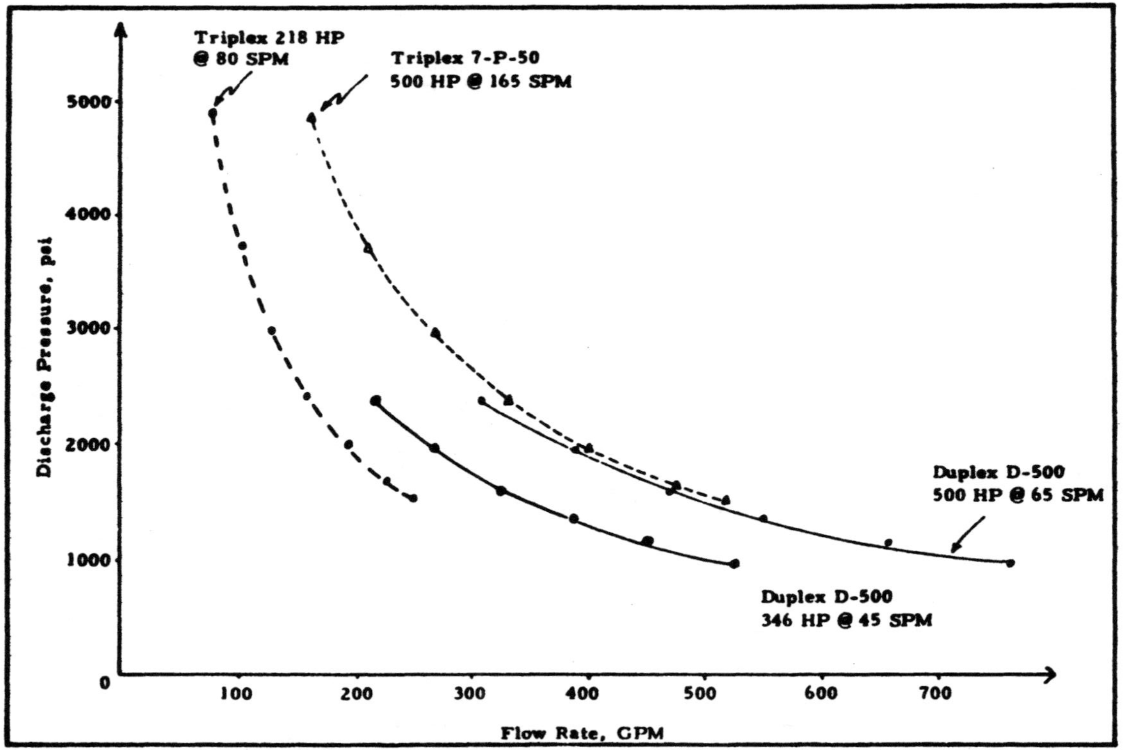

Figure 4.12 Comparison of pump performance between two types of pumps.

AVAILABLE HYDRAULIC HORSEPOWER

At any given operating condition, the power available is represented by the area of the rectangle under the pump performance curve:

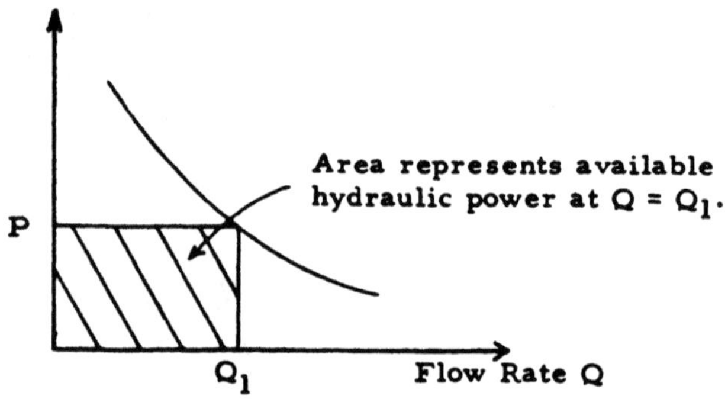

This available power will be used partly to overcome the frictional losses in the drillpipe and annulus and partly for cleaning the bottom of the hole and transporting the cuttings to the surface.

It is often advantageous to represent the pressure-flow relation for pump performance as a logarithmic plot, since from equation (3),

$$\text{Log HHP} + \text{Log } 1714 = \text{Log P} + \text{Log Q}$$

for a given input power, Log HHP + Log 1714 is constant and

$$\text{Log P} = \text{Constant} - \text{Log Q}$$

which plots as a straight line on log-log paper:

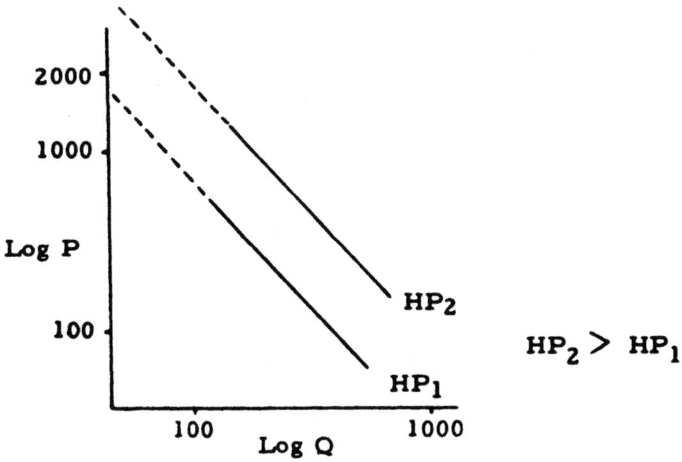

Since the intercept on the P axis is a function of the power level, a family of parallel lines of equal negative slope is obtained corresponding to the different power levels.

THE SUBSURFACE SYSTEM

Wellbore Hydraulics. The subsurface system is comprised of the drillpipe, drillcollars, bit, and annular space between the hole walls and the pipe. In general the drilling fluid flows into the drillpipe to the bit and back to the surface through the annular space. This is known as normal circulation, while the opposite flow path (down annulus, up pipe) is known as reverse circulation.

The wellbore geometry can be as simple as that of a constant diameter hole to the bottom of the well with sections of pipe of two different diameters, corresponding to the drillpipe and drillcollars. This is the case in shallow wells or during the drilling of surface hole in deep wells, as shown in figure 4.13.

As well depth increases, it often becomes necessary to set protective or intermediate casing strings, decrease the bit diameter, and use more complex drillstem assemblies designed to satisfy the strength requirements of the drillpipe and the

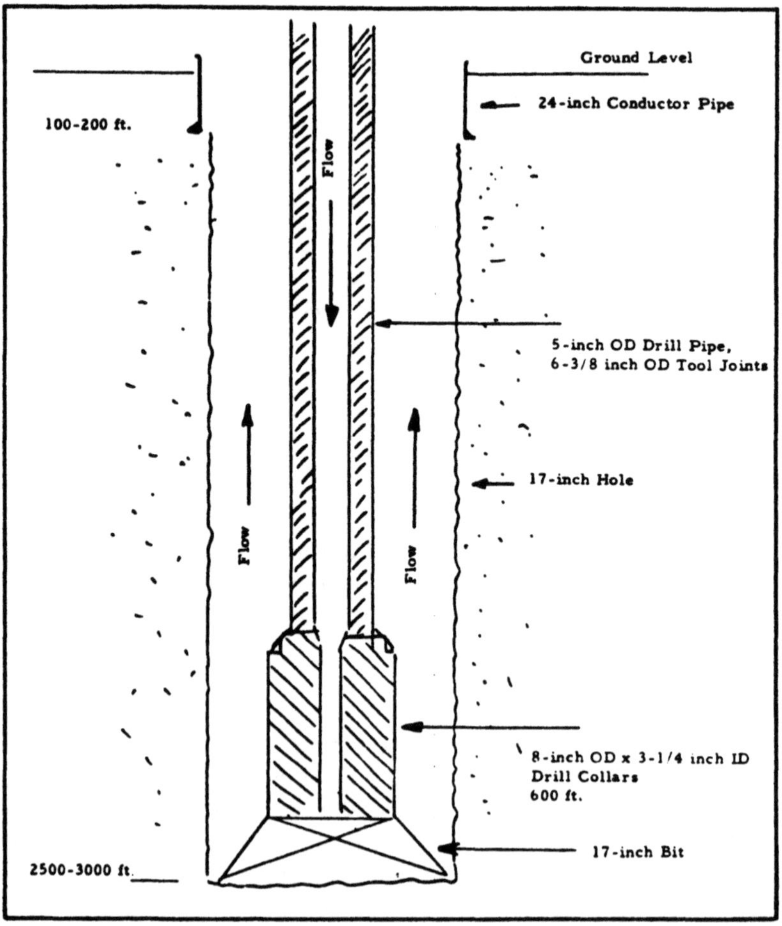

Figure 4.13 Drillstem configuration for surface hole drilling.

directional restrictions on hole deviation. This results in more complicated geometries with several sections of different hole diameters (corresponding either to open hole or casing of different sizes) and different pipe diameters (corresponding to tapered drillstrings and drillcollar sections). A typical combination of sizes is shown in figure 4.14. It should also be considered

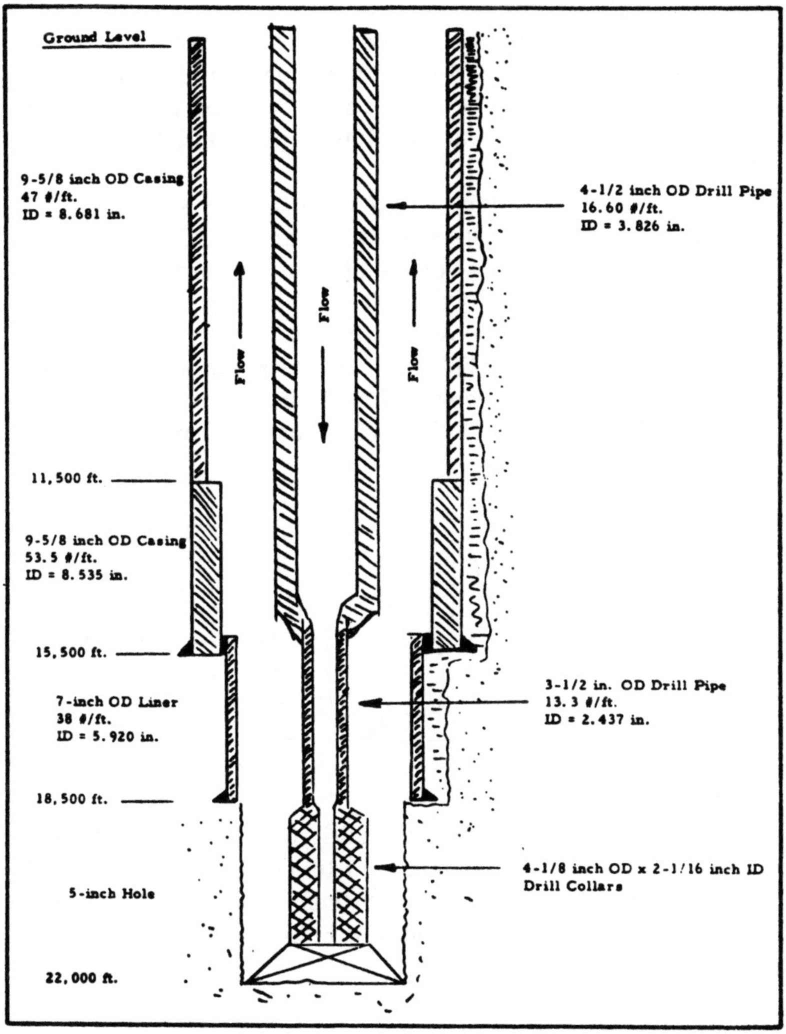

Figure 4.14 Drillstem configuration for deep hole drilling.

that the pipe's outside diameter is not constant, since every 30 feet there is an enlargement at the couplings. Also there may be drillpipe protectors spaced along the pipe to reduce pipe wear due to friction with the hole wall. In deep wells this may constitute a significant length (as much as 300 ft. per 10,000 ft. of pipe). This cannot be ignored.

Under normal conditions, the subsurface system is considered to be a closed system, so the volume of fluid returning to the surface is equal to the volume of fluid pumped into the drillpipe. (If the volume is less there is lost circulation; if it is more, there is a kick). Being a closed system, it is possible to view the flow column as pipes of different diameters placed in series and forming a U-tube with a constriction at the bottom corresponding to the drill bit (or bit nozzles). This is illustrated on the following drawing:

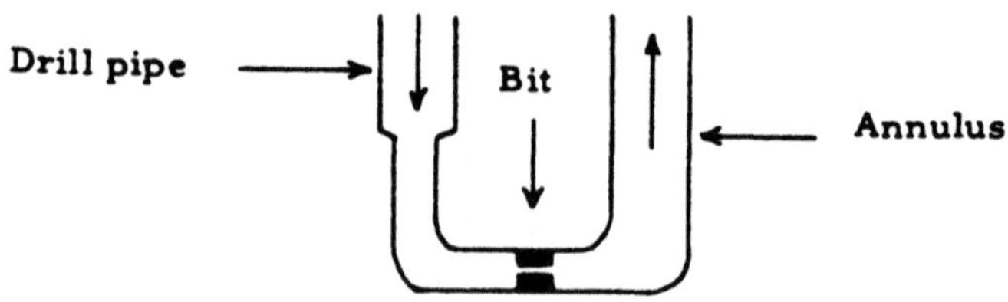

Under static conditions (no fluid circulation) the pressure distribution in the flow column as a function of depth depends only on the density of the fluid in the wellbore and is the same on either side of the pipe. The relationship between mud weight and pressure at a given depth is:

$$P = \frac{MW\ (\text{ppg})\ (0.433\ \text{psi/ft}) \times D\ (\text{feet})}{(8.33)\ (ppg)}$$

or

$$P = (0.052)\ (MW)\ (D)$$

where P is in psi.

MW is in ppg, and

D is in feet.

Figure 4.15 shows this relationship for different mud weights. The slope of each curve corresponds to the pressure gradient for the particular mud. It is generally assumed that the fluid is homogeneous and that there are no variations in density with depth. This may not be true in deep wells when using weighted fluids, since some settling of the Barite occurs when there is no circulation for a period of time.

Fluid circulation is established by increasing the pressure at the top of the drillpipe to the *circulating standpipe pressure,* P_s, corresponding to the flow rate desired. This is illustrated in figure 4.16. The pressure at any point in the flow column is the combination of the hydrostatic pressure, plus the standpipe pressure, less the sum of frictional losses to that point. In this manner, a complete pressure traverse can be obtained representing the pressure distribution in the flow column *for a given fluid and a given flow rate.* Several items should be pointed out with regard to this pressure distribution:

1. At a given flow rate and fluid properties, the pressure losses are a function of the wellbore geometry.
2. The pressure drop at the bit is significant and of the same order of magnitude as the pressure drop due to friction in the pipe.
3. The pressure at the bottom of the hole while circulating is greater than the hydrostatic pressure.
4. All the pressure available at the standpipe is expended in the flow column.

For the purposes of this discussion, table 4.4 shows the relationship between the static and flowing pressure distribution.

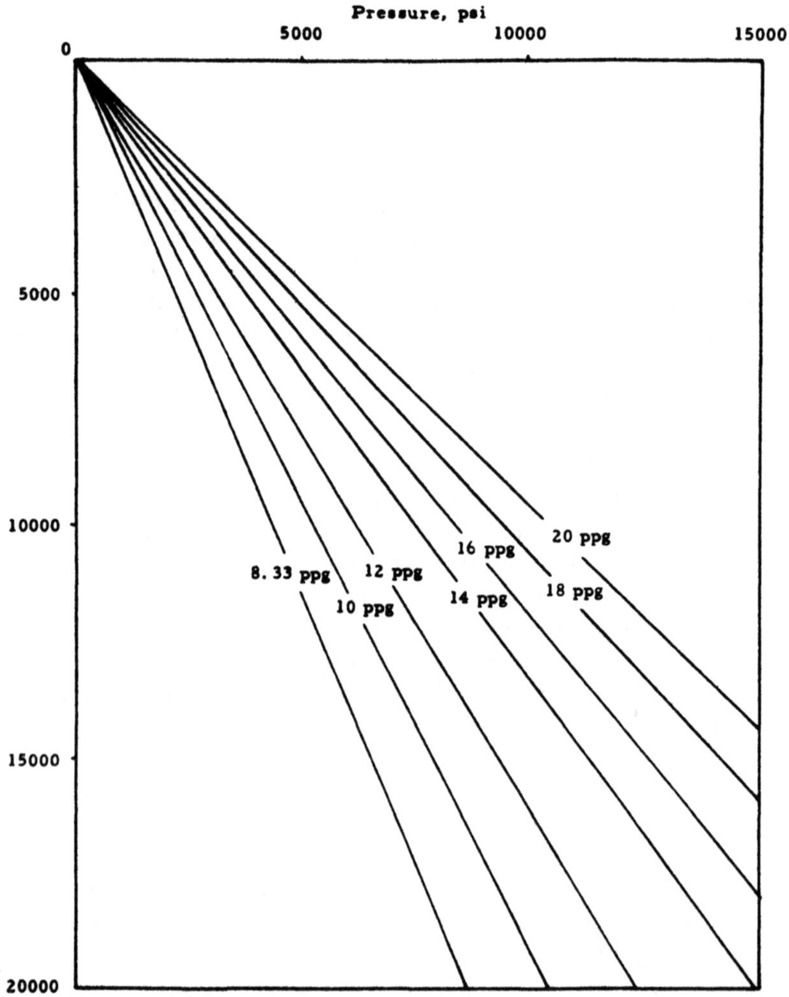

Figure 4.15 Relationship between mud weight and pressure at different depths.

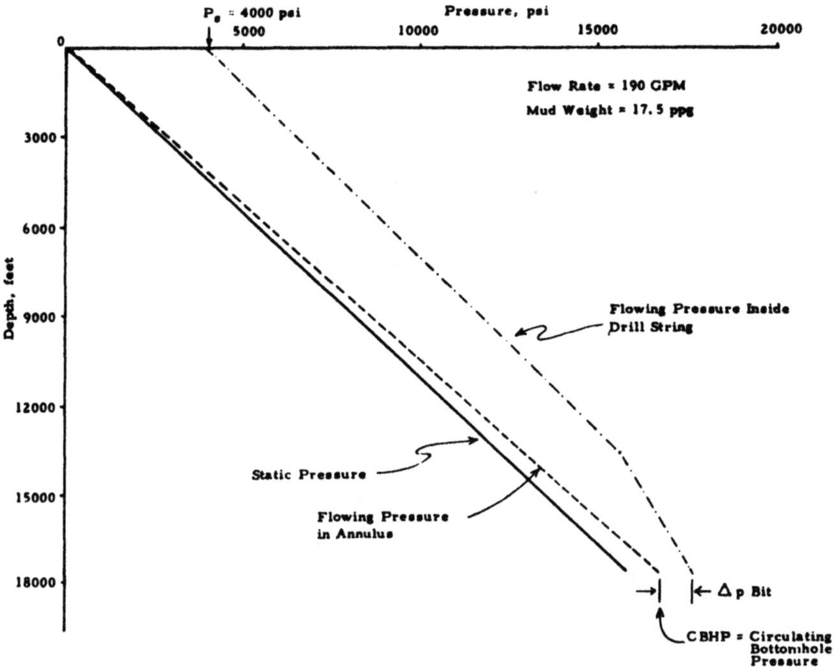

Figure 4.16 Plot of depth (ft) vs. pressure (psi).

Equivalent Circulating Density. As pointed out above, the pressure at the bottom of the hole while circulating (CBHP, circulating bottomhole pressure) is greater than the hydrostatic pressure developed by the fluid column. The difference corresponds to the frictional flow losses in the annulus, i.e., the excess pressure required at the bottom of the hole needed to flow the drilling fluid back to the surface.

Knowing the value of CBHP, it is possible to calculate the mud weight for a static column that would develop the same pressure. This mud weight is defined as equivalent circulating density (ECD):

$$\mathrm{ECD} = \frac{\mathrm{CBHP\ (psi)}}{\mathrm{D\ (ft)}} \times \frac{1}{0.052\ \mathrm{(psi/ft/ppg)}}$$

TABLE 4.4 FLOW COLUMN PRESSURE BALANCE

FLOW

Depth Ft.	Drill Pipe Static Pressure	Drill Pipe Friction Losses	Drill Pipe Pumping Pressure	Drill Pipe Net Pressure	Annulus Static Pressure	Annulus Friction Losses	Annulus Pumping Pressure	Annulus Net Pressure
0	0		4000	4000	0		0	0
		591				370		
11,582	10539		3409	13948	10,539		370	10909
		101				66		
13,568	12330		3308	15638	12,330		436	12766
		584				48		
15,447	14039		2724	16766	14,039		484	14523
		551				143		
17,221	15653		2173	17526	15,653		627	16280
		633				129		
18,137	16486		1540	18026	16,486		756	17242

Bit Pressure Drop = 784 psi

Standpipe Pressure = 4000 psi

Flow Rate = 190 gpm

Mud Weight = 17.5 ppg

Plastic Viscosity = 88 cp.

or, in the example:

$$\text{ECD} = \frac{17242}{18137} \times \frac{1}{0.052} = 18.3 \text{ ppg.}$$

Therefore, although the fluid density is only 17.5 ppg, when the fluid is circulated at 190 gpm the apparent density of the column increases to 18.3. (Intuitively the higher the flow rate, the greater will be the increase as will be shown.) This factor is extremely important if formations of low fracturing pressure have to be drilled since it may be the cause of lost circulation problems.

Considering only the dynamic pressure due to fluid circulation, that is, neglecting the hydrostatic pressure, the circulation pressure distribution is obtained as shown in figure 4.17. The pressure losses can be grouped into losses in the drill pipe (ΔP_P), losses across the bit (ΔP_b), and annuluar losses (ΔP_a). Their relative magnitudes will vary with the wellbore configuration, but in general the bit and pipe losses comprise about 80-90 percent of the total available pressure. The pressure loss gradients (p/unit length) vary inversely with the diameters of the pipe (or annular clearance).

A detailed analysis of the circulating losses becomes very important as the depth of the well increases since it is necessary to determine how the hydraulic energy is being utilized to insure an efficient drilling operation. Since the pressure losses are functions of the geometry, the fluid properties, and the flow rate, the detailed analysis requires a complete knowledge of these parameters as a function of depth.

Two cases can be considered: one from the standpoint of designing a program for a future drilling operation and to set guidelines for optimum equipment selection, the other from the standpoint of analyzing an existing system in order to improve its performance. In either case the basis for the solution is the definition of the volumetric requirements for the given condition, since this is necessary for the estimation of the pressure losses for a given wellbore geometry and fluid properties. In general it can be said that the volumetric requirement is set by the wellbore cleaning requirement, which is controlled by the transport

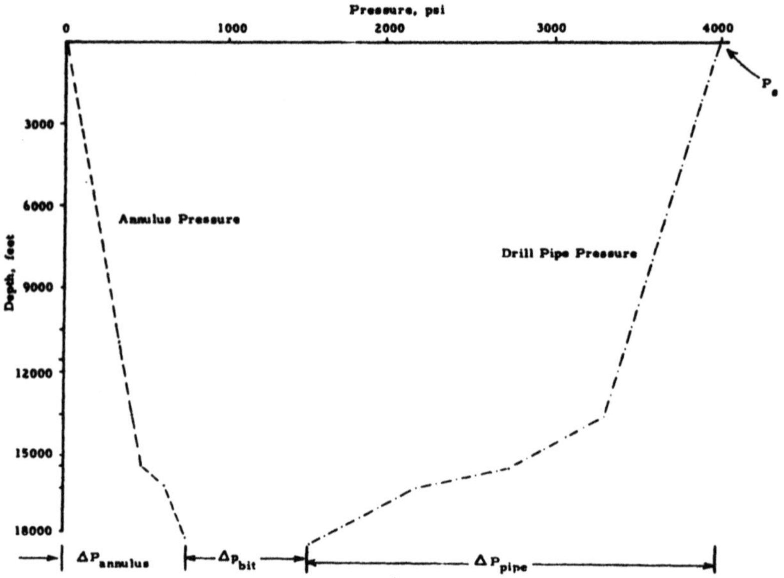

Figure 4.17 Circulation pressure distribution.

efficiency of the drilled cuttings. The pressure needs will be determined by the pressure losses in the flow column plus the power required to maintain the bottom of the hole free of chips (bottomhole cleaning). The problem can be focused in the following two questions:

(1) Given a hole size, a drillstring size, and a formation being drilled at a certain penetration rate (ft/hr), what volume of fluid will have to be pumped to move the rock cuttings (lbs/hr) at a transport rate that prevents the accumulation of cuttings in the well?

(2) At the flowrate for borehole cleaning, what pressure will be required at the surface to overcome frictional losses in the pipe and still have sufficient power at the bit to keep the bottom free of chips?

A possible method for the solution of this problem could be as follows:

a) For a given pipe-hole geometry, establish the flow rate necessary for proper borehole cleaning.
b) At this flow rate and for the existing fluid properties, establish the pressure losses in the pipe and annulus.
c) Establish the hydraulic power required for the efficient bottomhole cleaning and determing the equivalent pressure loss at the established flow rate.
d) Determine the total power required at the surface by adding the individual requirements of (b) and (c).

This procedure is repeated for each section of the well, wherever changes of hole or pipe size take place, and for the particular drilling fluid properties for that section. If the analysis is made for the preparation of a drilling program for a future well, it will be necessary to keep in mind any unusual problems or conditions that would impose restrictions on the possible choices. Appropriate hydraulic analysis has to be integrated into the total planning operation.

Whenever analyzing an existing system, the same procedure can be followed except that in this case there is an existing limit as to the available power at the surface, and the operating conditions (P and Q) have to be established so as to result in the best utilization of hydraulic energy for the drilling operation.

In either case, the following must be established:

a) Adequate flow rate for borehole cleaning.
b) Requirements for bottomhole cleaning (and bit cleaning).
c) Pressure losses in the system.

These three topics will be discussed in the following section.

DRILL CUTTING TRANSPORT

The buildup of cuttings in the annulus is detrimental to the overall drilling operation for several reasons:

1) An increase of cuttings concentration causes an increase in the density of the annular fluid, giving rise to lost circulation problems in areas of low fracture gradients.
2) A long residence of the cuttings in the annulus increases the amount of fine solids dispersed in the fluid due to the grinding action of the pipe.
3) In extreme cases, cuttings adhere to the drillstring, thereby increasing pressure losses and probability of pipe sticking.
4) Problems may arise caused by borehole instability.
5) Confusion and even erroneous conclusions as to lithology may result from cuttings analysis, with attendent problems and improper interpretations.

Several investigators have studied the problem of determining the volume and velocity of the drilling fluid needed to keep the wellbore "clean" and with a low concentration of drill cuttings. Early results indicated that this problem was equivalent to determining the velocity with which the solid particles settled in the fluid (slip velocity) under the action of gravity, and then selecting a volumetric flow rate that for the particular annular area would result in an annular flow velocity exceeding the slip velocity. In this manner the cuttings exhibit a net upward velocity relative to the wellbore:

$$V_c = V_a - V_s$$

where

V_c = velocity of cutting relative to the wellbore

V_a = annular fluid velocity

V_s = slip velocity of cutting in fluid

It was found that the slip velocity is a function of the flow regimen (laminar or turbulent), of the properties of the fluid such as

viscosity and density, of the properties of the solid particles, and of the shape and size of the cuttings. This greatly complicates the calculation of slip velocity in the case of drill cuttings due to the variation of shape, size, and composition of the rock chips as well as the variation of the fluid properties with depth, temperature, and flow conditions. Nevertheless relationships were developed that approximate the actual slip velocity of flat cuttings.
For laminar flow:

$$V_s = \frac{57.5d^2\ (\rho_s - \rho_e)}{\mu}$$

and for turbulent flow:

$$V_s = (133)\ \left(\frac{t_c}{d_c}\right)^{1/2}\ \left(\frac{d_c\ (\rho_c - \rho_e)}{\rho_e}\right)^{1/2}$$

where

d_c = cutting diameter (in.)

t_c = cutting thickness (in.)

ρ_e = liquid density (ppg)

ρ_c = cutting density (ppg)

μ = liquid viscosity (cp)

V_s = slip velocity (ft/min)

Early experiments indicated that in the specific case of annular flow the following generalizations could be made:

1) Turbulent flow in the annulus improved cutting transport.
2) Low viscosity was desirable because it increased turbulence.
3) Pipe rotation increased transport efficiency since it

causes the cuttings to remain in the center of the annulus.

4) Annular fluid velocities of 100-125 ft/min are needed to maintain a clean wellbore.

Based on these findings and taking into account the effect of fluid density (increasing buoyancy with increased density) and annular clearance on the onset of turbulence, a generalized relationship was developed to establish guidelines for flow-rate requirements. A typical relationship is shown in figure 4.18. However, problems were encountered in some areas due to fluid erosion of the wellbore walls and high pressure losses in the annulus. With increasing well depth requiring multiple casing strings and large surface holes, and with the increase of wells drilled offshore from floating vessels using large diameter risers, it became apparent that these guidelines imposed the need for excessive hydraulic power. This prompted further experimental studies which indicated that:

a) The maximum slip velocity of single cuttings is 100 to 110 ft/min in low viscosity, low density drilling fluid.
b) The slip velocity is an inverse function of fluid viscosity. In the case of non-Newtonian fluids, yield point is the property that gives the best correlation.
c) The slip velocity decreases sharply for viscosity above a certain threshold value (see figure 4.19).
d) Slip velocity decreases with increasing fluid density due to the increase in buoyancy.
e) A review of field cases indicates that no problems were encountered during drilling operations whenever the cuttings concentration in the annulus was below 5% by volume, suggesting that for best operation the rate of cutting removal has to increase as drilling rate increases.

Recent work by Ziedler (1972) in a 65-ft., 8 1/2-in. casing with 4 1/2-in. OD drillpipe system indicates that the concentration of cuttings has a net effect on transport efficiency, with an increase in efficiency caused by increases in concentration. This

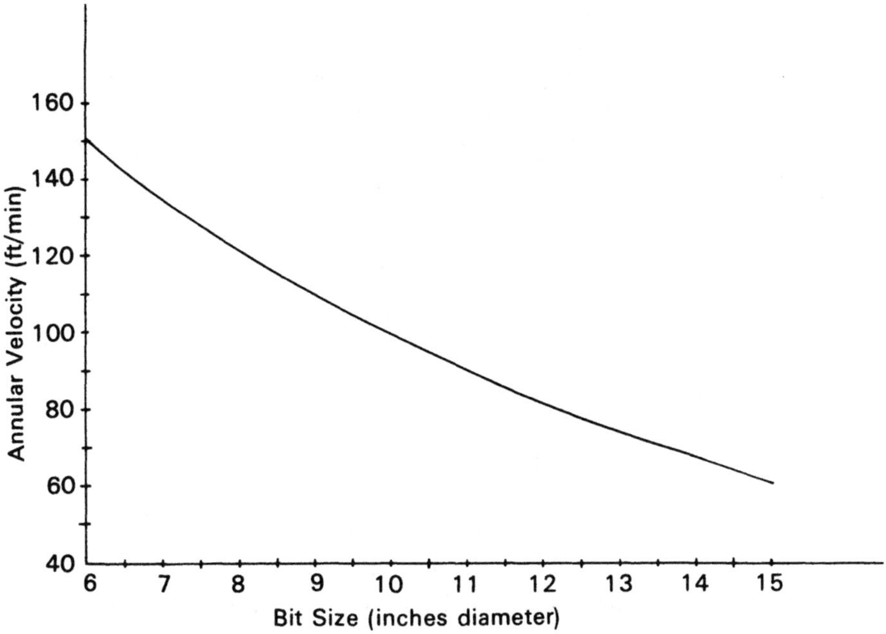

Figure 4.18 Normal minimum annular velocity with 9.2 lb/gal fluid.

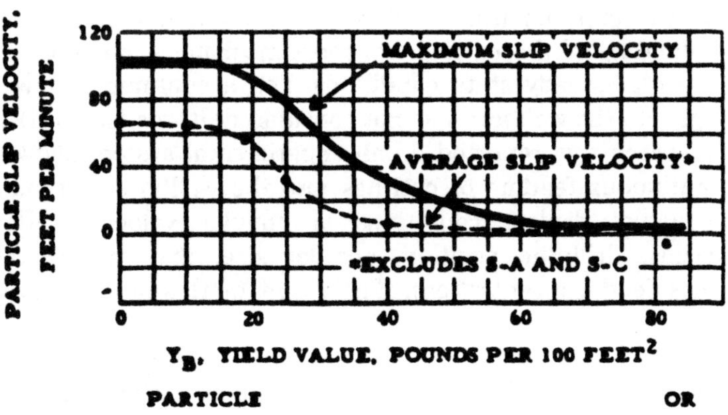

Figure 4.19 Particle slip velocity vs. yield value for all particles.

suggests that single cutting slip velocity does not necessarily reflect the transport efficiency of the system, but that interaction between the particles has to be included in the analysis. This experimental study resulted in the definition of a recovery fraction, corresponding to the percentage of particles transported to the top of the annulus from an initial quantity of particles at the bottom:

$$f_c = 1 - e^{-0.0273 \left(\frac{V - V_s}{V_s}\right)^{0.96} \cdot N^{0.29}}$$

where V_s is the single particle slip velocity

V is the velocity of the liquid

N is the particle concentration

This equation indicates that the recovery fraction is never 100% and that recovery increases as the velocity of the liquid increases (for a given liquid-particle combination) and increases slightly as the concentration of solids increases.

These experimental studies contributed greatly to the understanding of the cutting transport mechanism but they did not consider steady state conditions, i.e., a continuous flow of cuttings into the wellbore was not considered. It was only in 1973 that research was reported on an experimental system allowing the continuous feeding of cuttings into the wellbore and simulating actual drilling conditions of annular velocities, fluid properties, drillpipe position, annular size, and drilling rate (cuttings feed concentration). The result of this study showed that:

1) The most important factors controlling cutting transport are *annular velocity* and *rheological properties* of the fluid.
2) Annular velocities of 50 ft/min provide satisfactory cutting transport in typical drilling muds.

3) Cutting transport efficiency increases as *viscosity* increases.
4) Transport efficiency is less than predicted by theoretical relationships. This was attributed to the shear-thinning properties of the liquids used.
5) Cutting size and fluid density have moderate influence on transport efficiency. Increases were observed with decrease in cutting size and increase in density of the fluid.
6) Hole size (or casing), drillpipe rotation, and drilling rate have only a slight effect on cutting transport.

The study is based on the determination of the transport ratio, which is defined from the relationship:

$$V_c = V_a - V_s$$

by dividing both sides by V_a,

$$\text{transport ratio} = \text{TR} = \frac{V_c}{V_a} = 1 - \frac{V_s}{V_a}$$

which corresponds to the fractional velocity of the cutting with respect to fluid velocity.

Considering a given penetration rate and hole size, it is possible to determine the volume of cuttings entering the wellbore per unit time:

$$C_v = R \times A$$

where A = area of bottom of hole which can be expressed as a concentration of cuttings in the volume of fluid being pumped in the system:

$$C_f = \frac{C_v}{Q}$$

where

C_f = cutting feed concentration

C_v = volume of cuttings entering wellbore per unit time

Q = volume of fluid pumped per unit time

If the transport ratio were equal to unity, indicating no slip velocity between cuttings and fluid, the cutting concentration at any point in the annulus would be equal to the feed concentration, c_f. However, in general, the transport ratio is less than unity, $TR = 1 - V_s/V_a$, resulting in an increase in cutting concentration in the annulus. At steady state conditions, the concentration of cuttings in the wellbore stabilizes to a certain value depending on the transport ratio. For example, assuming a feed concentration of 2% and a transport ratio $V_c/V_a = 0.5$ results in a steady state concentration of 4% in the annulus. Therefore, the concentration in the annulus C_a, is given by:

$$C_a = C_f \cdot V_a/V_c = C_f/TR\,.$$

Figure 4.20 shows typical transport ratio as a function of annular velocity. It is interesting to note that although transport ratio increases with increasing annular velocity, it does not reach 100%. Even at low annular velocity it is possible to obtain transport coefficients in excess of 50% with normal drilling fluids, as shown by figure 4.21. For very low viscosity fluids it is necessary to increase the annular velocity to a minimum value before appreciable transport is established. This minimum value greatly decreases as soon as the viscosity of the liquid is increased slightly.

Combining these results with the earlier findings that wellbore cleaning problems can be avoided by keeping the cuttings concentration below a certain limit, such as 5%, it is possible to establish an annular velocity for a given drilling rate that will result in this condition being satisfied. The cutting feed concentration is given by:

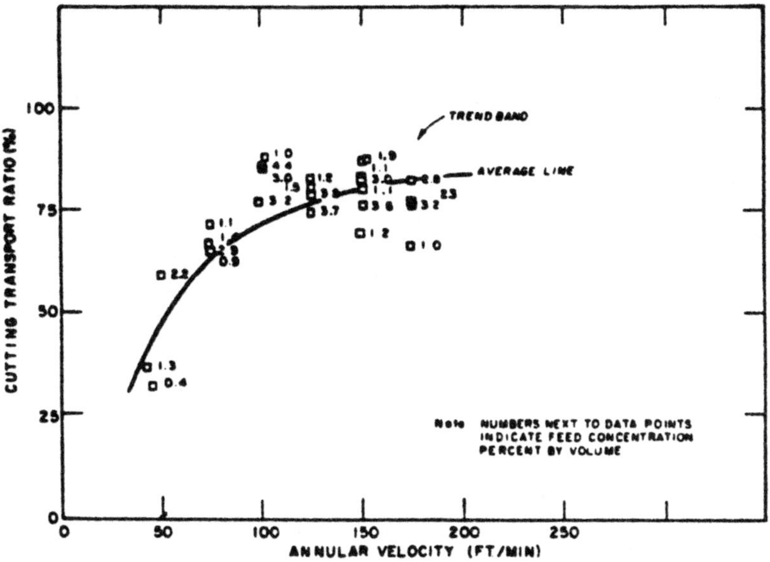

Figure 4.20 Typical cutting transport results (medium cutting, no rotation, 8 × 4-in. annulus, 12 ppg mud).

$$C_f = \frac{R \times A}{Q}$$

and the annular concentration

$$C_a - C_f \frac{V_a}{V_c}$$

where

$$V_a = \frac{Q}{A_a} = \frac{Q}{A - A_p}$$

where A_p = area of pipe O.D. therefore

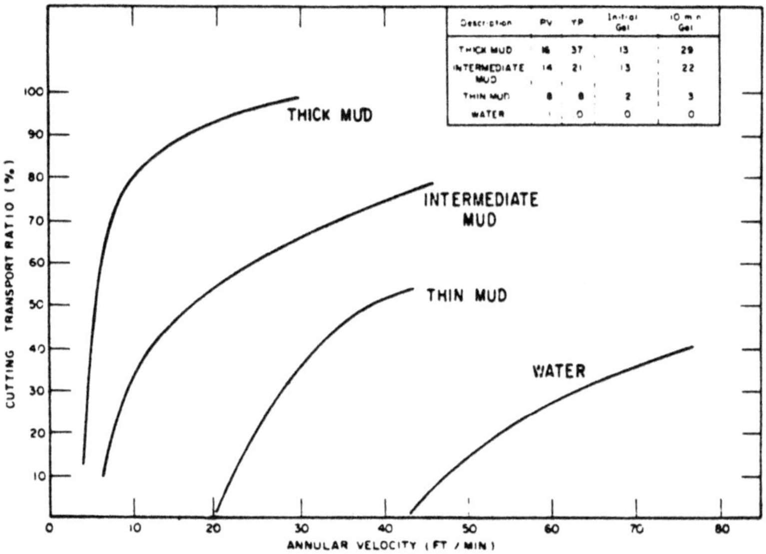

Figure 4.21 Cutting removal at low annular velocities (medium cutting, no rotation, 12 × 3 1/2-in. annulus, 12 ppg mud).

$$C_a = C_f \cdot \frac{Q}{A - A_p} \cdot \frac{1}{V_c}$$

$$C_a = R \cdot \frac{1}{1 - A_p/A} \cdot \frac{1}{V_c}$$

but

$$V_c = (TR)\ V_a$$

or

$$C_a = \frac{R}{(TR) V_a} \cdot \frac{1}{1 - A_p/A}$$

For a desired cutting concentration C_a

$$V_a \geq \frac{R}{(TR)C_a}\left(\frac{1}{1-(D_p/Dh)^2}\right)$$

where

R = drilling rate (ft/min)

TR = transport ratio (%)

C_a = annular cutting concentration (%)

D_p = pipe O.D. (in)

Dh = hole I.D. (in)

Conversely for a given annular velocity the maximum drilling rate that can be tolerated that will result in a cutting concentration in the annulus equal to or less than C_a is given by:

$$R \leq (V_a)\ (TR)\ (C_a)\ \left(1-\left(\frac{D_p}{Dh}\right)^2\right).$$

Table 4.5 was calculated assuming the case of a thin mud, 12-in. ID hole, and 4-inch pipe.

TABLE 4.5 ANNULAR CUTTING CONCENTRATION 5%

Max. rate (ft/hr)	Annular veloc. (ft/min)	Transport ratio
0	20	0
13.4	25	0.20
28	30	0.35
54	40	0.50
75	50	0.55
130	70	0.70
200	100	0.72
303	150	0.75
405	200	0.76

Figure 4.22 shows these results and curves similarly calculated for the case of water and thick mud.

Similar curves can be calculated for other combinations of wellbore size and can be used as guidelines for choosing appropriate annular velocities as a function of penetration rate to maintain the concentration of cuttings in the annulus below an established minimum value.

In summary it can be stated that:

a) Wellbore cleaning can be established at low drilling fluid velocities for moderate to low viscosities.
b) A fluid velocity at a given drilling rate will establish a certain cutting concentration in the annulus, which should be below a certain minimum value.
c) For a given drilling rate, increasing annular velocity above this point does not result in improved wellbore cleaning.
d) Guidelines based on 120 ft/min minimum annular velocity (or higher) generally result in excessive flow rates for drilling rates of 120 ft/hr or less. However these high velocities may be required for very high penetration rates or very thin fluids.

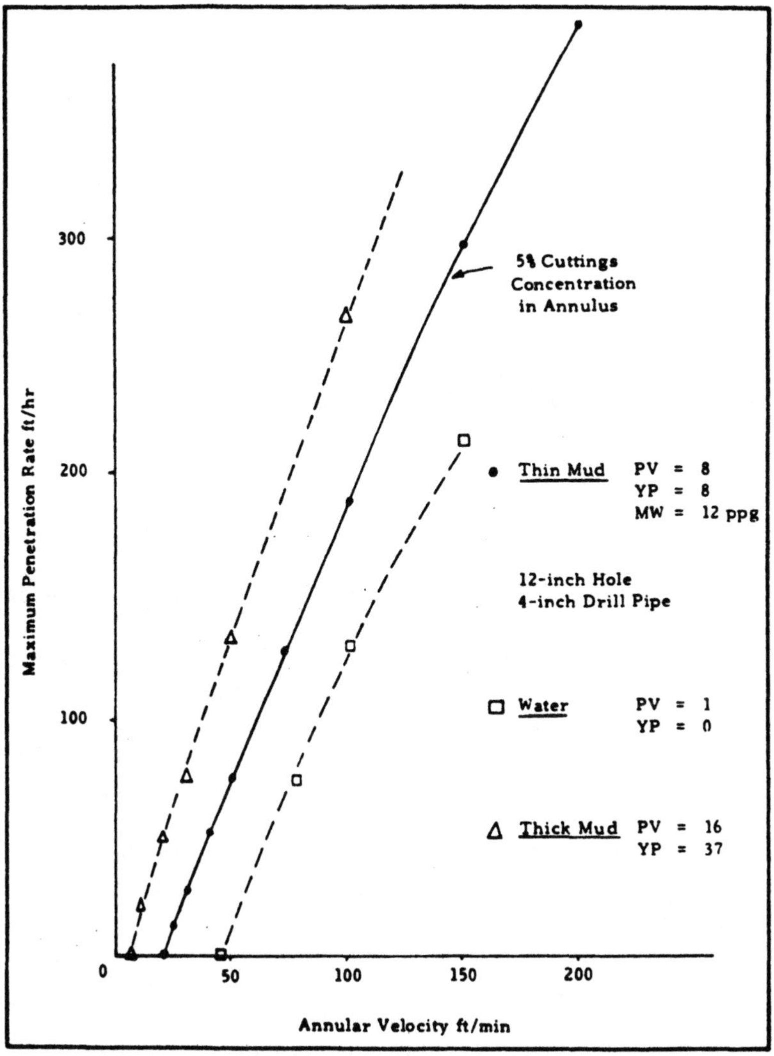

Figure 4.22 Plot of maximum penetration rate (ft/hr) vs. annular velocity (ft/min).

5
Drillstem Testing

A so-called drillstem test is more aptly described under the name of *formation test* and is generally performed during the course of drilling. Special formation test equipment is mounted on the end of the drillstring and lowered into the hole to a point above or adjacent to the horizon to be tested. Wireline formation testers are also used, especially where a series of formations are to be tested consecutively, and are generally run in a cased hole.

Drillstem tests are made to ascertain the potential productivity of a penetrated zone, to access formation damage, to determine native reservoir pressures, and to obtain fluid samples (surface and/or subsurface) without cementing casing or removing drilling fluid from the hole. Such tests can also be used to confirm the effectiveness of water shutoffs and to determine the capability of perforations to admit fluids freely to the hole.

Drillstem testing, its planning, implementation, and analysis, is a specialized field. It can also be quite costly and at times even dangerous to equipment and personnel. Therefore, expert service company personnel are invariably called in to assist in the planning, to provide the tools and associated equipment, and to supervise the test program.

5.1 Analysis of Need

The need and justification for a drillstem test or a program of tests is best indicated by the quantitative and qualitative information that can be derived from a successfully run test. Quantitative data that can be obtained from the test of a potentially productive formation include the formation pressure, formation damage or skin-factor, and formation productivity index. Qualitative and semiquantitative information may include the type of fluid to be produced, the presence or absence of *nearby* faults and fluid contacts, the indication of an on-site fracture system, and even the productive sand volume in the case of small, closed reservoirs. In short, if surface and/or bottomhole fluid samples are also taken during the test, all of the various geological and engineering data and information obtainable from the testing and sampling of a stabilized producing well can be obtained from a drillstem test. With modern equipment, test procedures, and analysis techniques, these data can be as valid and reliable as those obtained under the more closely controlled conditions of production well tests.

Other uses for drillstem tests, referred to previously, include the confirmation that a casing or squeeze job for water shutoff has been successful and the identification of approximate reservoir gas–oil and oil–water levels for the purpose of selecting an appropriate perforation interval.

Knowing the information that can be obtained and the current and future requirements for various data and information, one can define the purpose and goals of a drillstem test program. This must then be balanced with the costs and risks involved to determine whether the need outweighs the negative aspects and whether there is a reasonable chance for a successful testing program.

The problems and dangers that can be encountered in drillstem testing are discussed more fully later, but they can be divided into three broad categories:

1. disasters that would entail loss of the rig or hole and/or constitute serious personnel hazards;

2. major expenses that would result from loss of tools, fishing jobs, inordinately long rig downtime, etc., and
3. minor expenses that would result from test problems, such as packers failing to seat, improper choke sizes, tools failing to open, clock stoppage, etc.

If disaster-type risks are present, the test usually should not be run and, at the least, high-level safety precautions must be taken. Major-expense risks must be seriously weighed against the potential costs, the probability of occurrence, and the value of the information to be derived. Minor-expense risks can be partially avoided by proper planning, but even then some problems will be incurred with many tests and test assemblies. The choice of taking such risks and the estimated corrective expenses should be handled on a statistical basis.

When weighing risks of any of the three types, it is important to evaluate past problems and accidents on the basis of those failures which occurred from risks that were intelligently taken and those that resulted from poor practices and from poor planning. Then, if the need for and value of the data obtained exceeds the probable costs, the drillstem test program can be justified.

5.2 Equipment

Basically, a drillstem test involves the measurement of bottom-hole pressures with the formation to be tested alternatively closed-in and open to flow to the surface. Therefore, the equipment in the overall assembly consists of the pressure-recording device, the flow-control valves, one or more gland-type packers, and various other safety and control mechanisms. Surface equipment may also include pressure and flow measuring and control devices, along with necessary manifolds, tanks, separators, burners, etc.

Tests in shallow low-pressure wells are made with comparatively simple assemblies, but those in deeper wells, with higher pressures and temperatures and relatively longer intervals of exposed hole, require various types of more complex and complete assemblies. Four sets of tools and drillstem test assemblies are shown in figures 5.1 and 5.2 for typical deep-well (10,000 to 20,000 + ft) testing. The equipment shown in figure 5.1 are two types available from Halliburton Company, and the assemblies illustrated in figure 5.2 are those of Johnston Testers, the inventors of drillstem test equipment in 1926. Other similar assemblies are available from other service companies.

Figure 5.1(a) illustrates the typical downhole assembly for a single-packer test in an open hole. At the bottom of the assembly is a bourdon-tube clock-driven chart-type pressure recorder with a maximum-recording thermometer. Immediately above is the perforated anchor through which produced fluids pass into the drillstring. Above this are a safety joint and an expanding-shoe packer assembly that seals off the test zone below from the annular mud column above. This is topped by another safety joint, a set of hydraulic jars, and a second pressure recorder. Above this recorder, the hydrospring tester operates by weight and pipe rotation to open and close the tester and the bypass ports. The dual closed-in pressure valve, with reverse circulation ports, permits two cycles of flow and closed-in pressures to be taken. The ports can then be opened by further rotation to reverse the produced fluid out of the drillpipe. Above the CIP valve is the handling sub and downhole choke assembly for control of test flow rates and better well control when high pressures are encountered. Then, at an appropriate location in the drillpipe above the assembly, an impact reverse sub is often installed for further well safety.

The various bypass ports are used primarily to ease the passage of the tester assembly down the hole, minimizing the surging and swabbing effects. Safety joints are made to shear in tension or with strong rotative forces to release from stuck tools the free portion of the string. They are constructed to transmit torque in either direction and are unaffected by inertia, vibration, or rotating out of the hole since they require some type of positive mechanical manipulation to separate the two sections, depending upon the design.

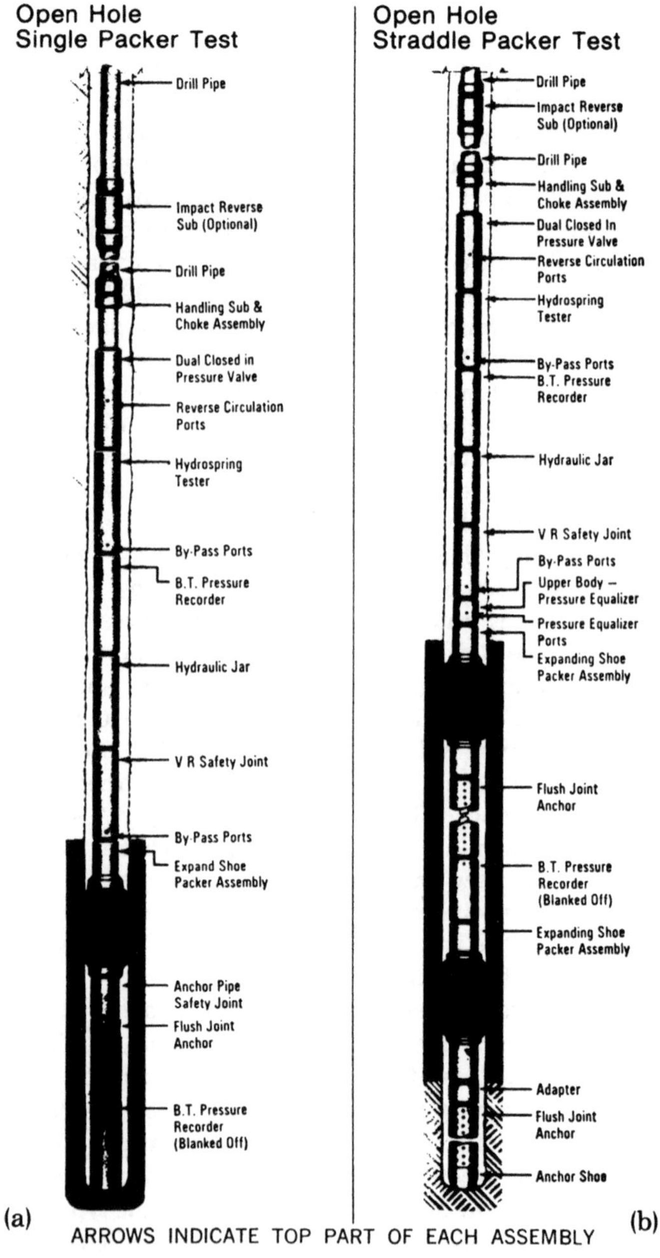

Figure 5.1 Typical DST assemblies. (Courtesy of Halliburton Services)

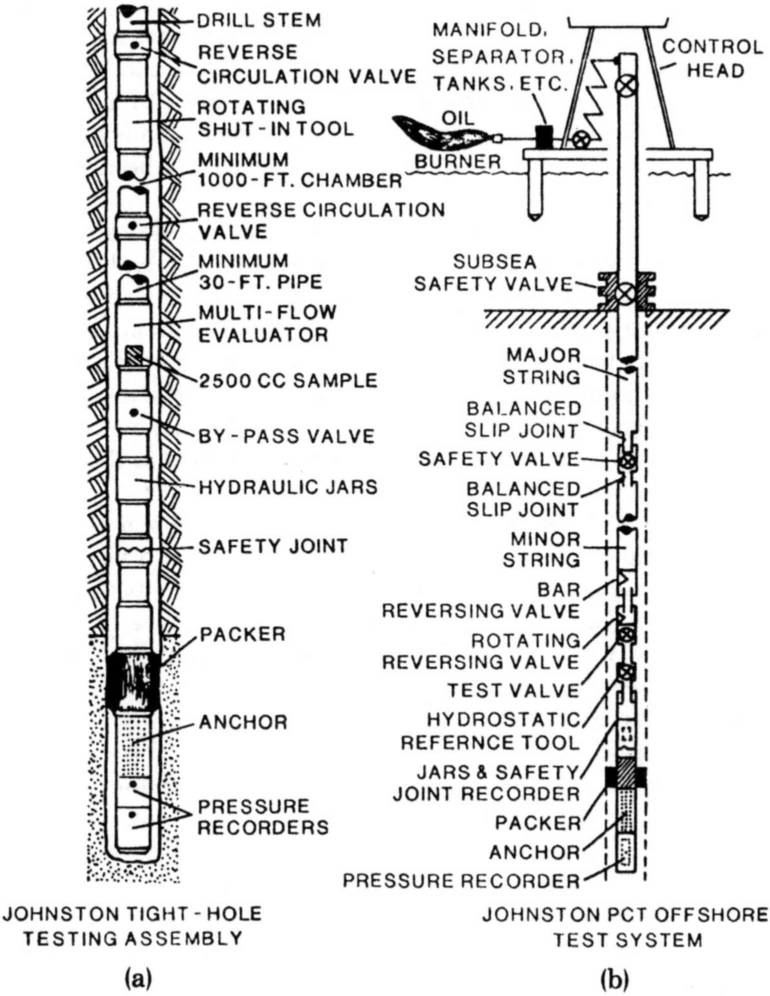

Figure 5.2 Typical DST assemblies. (Courtesy of Johnston-Macco)

The assembly of figure 5.1(b) is essentially the same as that of figure 5.1(a), except that it includes two packers and an additional anchor joint. With these tools, packers can be spaced to seal off both above and below the test interval to make a so-called straddle test. In addition to routine formation testing, this type of assembly is often used to test cement jobs and water shutoffs inside casing.

Figure 5.2 illustrates a test assembly that also includes a bottomhole sample chamber of 2500 cc capacity and a typical offshore test assembly with a seafloor safety valve and the necessary rig floor equipment. In the first system, so-called Johnston tight-hole testing assembly, two pressure recorders are run at the bottom of the string, and the opening and closing of the string is accomplished by means of a multiflow evaluator. Upon final closing, a sample of fluid is trapped and held in the sample chamber at the final build-up pressure.

The multiflow evaluator permits the opening and closing of the flow valve simply by raising and lowering the drill string to apply tension and then compression above the packer. Any number of cycles can be run. The valve opens on a time-delay mechanism when the drillstem is lowered against the packer, thereby preventing surging of fluids and pressures. Raising and lowering the pipe again closes the valve.

The rig-floor equipment noted in figure 5.2(b) includes a control head on top of the test string, a manifold, gas–oil separator, gauging and monitoring equipment, tankage, and a burner to dispose of excess produced fluids. The swivel-type control head with a double manifold permits safe changing of surface chokes and is often equipped with a bar-dropper assembly to permit the introduction of a go-devil to operate certain types of reverse-circulation valves.

It should be noted that the four assemblies of figures 5.1 and 5.2 are typical for many test operations but that many other specific assemblies of these tools and others are also used. All of the various types of equipment are generally available from the example companies as well as from several other companies that offer well-test services. Those shown here are included only for illustrative and explanation purposes.

5.3 Procedures

In a common drillstem test, the packer is first set against a competent formation, above and as near as practical to the zone of interest. The tester valves are then opened. Fluid from a permeable test interval will enter the ports of the test tools and rise into the drillpipe. The height and rate of the rising fluid will depend on the effective productivity and pressure of the formation and the weight of the column of fluid carried in the drillstring, either drilling mud or a water or inert-gas (nitrogen) cushion. The produced fluid will reach the surface and continue to flow if all conditions are favorable.

At the surface, a blow of "air" will usually be noted first from the open end of the system, with the rate of displacement being a measure of the rate of fluid rise in the drillpipe. If this fluid continues to rise, this blow will be followed by flows of drilling fluid (or cushion), gas-cut mud, gas, water, and/or oil. If the formation pressure is not great enough to lift the reservoir fluids to the surface and sustain a rate of flow, a pressure equilibrium will be obtained within the drillstring and all flow will cease. At this time or after a desired time of surface flow at one or more rates, the valves of the tester are closed and a circulating valve is opened to permit annular drilling fluid to fill the hole below the packer once again and equalize the pressure. The packer is then released and the drillstring is removed from the hole and uncoupled in "stands" with the trapped fluid inside. Thus, it may be found that fluids, such as gas, drilling mud, oil, and formation water are in the pipe. The number of "stands" or "joints" of each fluid is recorded and converted to barrels or cubic meters per hour as a basis for estimating the formation productivity. Samples of produced fluid can be captured and retained for analysis.

During the test, the surface assembly is designed to carry produced fluids to suitable storage, trapping, or disposal areas at some safe distance from the drilling rig. Usually this assembly will also be equipped with a control device or choke to restrict the rate of flow and protect the system from high pressures. Rate-metering equipment will also be included. Therefore, part of the test records include a description, both qualitative and quantitative, of the fluid production as a function of time. For example, gas flows may be described as: weak, medium, or strong blow of gas—15 minutes. Flow rates may be reported as: oil flow of 25 bbl/hr through $\frac{5}{16}$-in. choke. It is necessary that as much qualitative and quantitative data be recorded as can be obtained or estimated to serve as a basis for test interpretation.

The overall drillstem test may consist of only one opening and closing of the test tool, one flow rate with one pressure drawdown, and one pressure buildup to final static pressure. On the other hand, the test may include a series of openings and closings with several drawdowns and buildups with the same or different choke settings for each cycle or with the flow rate varied during one or all cycles.

All such procedures are generally included in the overall test planning in order to obtain the definite information desired and/or required for a particular well, zone, and reservoir.

To obtain a successful test, it is necessary that the packer or packers seal effectively, that all equipment functions properly, and that test procedures are planned and carried out so as to permit stabilized conditions to be obtained in each portion of each test cycle. In the absence of any expertise to the contrary, certain rules of thumb have been proposed for various test operations, and these can be used as a starting point for proper planning:

1. The initial shut-in period should be at least 60 minutes in length. A study of thousands of tests showed that only 50% of the wells reached a maximum static pressure in 30 minutes, 75% in 45 minutes, and 92% in 60 minutes. In the last case, the other 8% were in tight formations and generally were noncommercial.
2. One-to-two hours of a good-to-strong blow will usually yield satisfactory drawdown data; otherwise, when the blow dies, the tool can be shut in for build up.
3. Shut-in time should never be less than 30 minutes. For a well with a good flow to the surface, it should be at least half the flowing time; for an average well flow, it should be equal to the flowing time; and for a well with poor flow

characteristics, it should be at least twice the flowing time.

The most common and aggravating mistake in drillstem tests is to fail to wait an adequate time before changing regimens. This includes the failure to obtain stabilized initial and final shut-in pressures and flowing pressures at each choke setting. If two or three flow rates are desired during one or more flow cycles, but hole conditions preclude leaving the tools in the hole for several hours, it is better to obtain data from only one flow rate and one cycle under stabilized conditions than to obtain only partial data from a series of rates.

In planning and conducting a test program, it is necessary to remember that each test is as individualistic as the formation being tested and that specific local experience is the only means by which a complex series of test procedures can be designed with any real degree of confidence.

5.4 Use of Data

The data from a drillstem test consists of all the qualitative, quantitative, and procedural information recorded at the surface and the pressure charts retrieved from the bottomhole test assembly. All of these data are necessary for complete and proper interpretation and use of the test results.

It is beyond the scope of this book to present a detailed comprehensive discussion of all of the various data and interpretations obtainable from drillstem test records and information. Therefore, only the more commonly derived qualitative information and quantitative data will be presented. Additional, more specialized information is available in the manuals prepared by the various service companies and in the petroleum engineering technical literature.

Figure 5.3 is an example of a typical, but idealized, drillstem test pressure chart for a test with one flow and one final buildup.

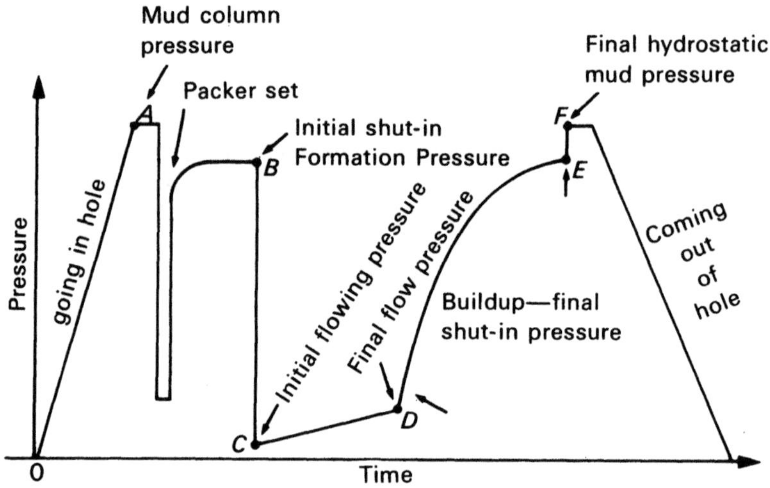

Figure 5.3 Idealized drillstem test pressure record.

This chart is a mechanical record of pressure (the ordinate) versus time (the abscissa).

The sloped line from zero to point A represents the increasing submergence pressure of the recorder as the test assembly was lowered into the hole, and the pressure at A is the initial static mud column pressure. Pressure B is the initial shut-in formation pressure after the packer was set, isolating the test interval. When the tool was opened, with little or no cushion in this case, the pressure C is the initial flowing pressure, and D is the final flowing pressure when the tool closed. The curve from D to E is the pressure build-up record to the final shut-in formation pressure at E. Then, F is the final hydrostatic mud pressure, and the remainder of the record is the pressure trace as the test assembly is removed from the hole.

Analysis of an actual chart record can indicate any of the possible types of mechanical problems that may have been encountered during the test and even the general quality of the test interval. The former is especially true when two pressure records are available from recorders located above and below the packer. Figure 5.4 shows the records obtained from a satisfactory single-flow test with two pressure recorders. In this case the bottom gauge was blanked off whereas the top gauge was not. The

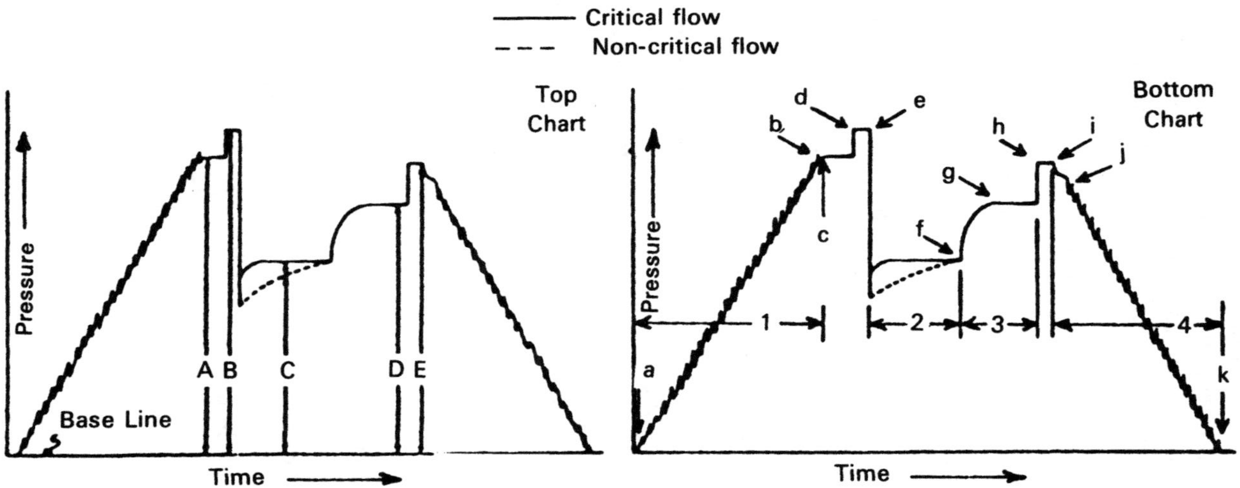

Figure 5.4 Dual-chart pressures and events. (In Black, WM, A review of drillstem testing techniques and analysis, *JPT*, page 21, June 1956. Courtesy of SPE-AIME, © 1956).

notations on these records indicate the different pressures and events during the test and can be used for reference in studying figures 5.5 and 5.6, which show representative records of test problems and different formation and flow characteristics, respectively.

The notations used in figure 5.4 are as follows:

Pressures:

A = initial hydrostatic mud pressure

B = packer squeeze pressure

C = average flowing pressure

D = final shut-in pressure

E = final mud pressure

$D - C$ = drawdown pressure

Times:

1 = running in

2 = flow period

3 = shut-in period

4 = time pulling out

Events:

a = first stand in

b = last stand in

c = tools on bottom

d = packer set

e = tool opened

f = tool closed

g = buildup complete

h = equalizing valve opened

i = packer unseated

j = first stand out

k = last stand out

Also, in figure 5.4, the solid line during the flow period

indicates a critical flow regimen, and the broken line indicates noncritical flow.

The charts in figure 5.5 show the results obtained from conditions that can be called mechanical failures which result in unsuccessful tests. Reference to the typical test-tool assembly and the sequence of test events will readily explain the pressure–time record of each chart.

Note that here the solid line represents the record from the top chart or both top and bottom chart, and the dashed line represents the bottom-chart record only.

The charts shown in figure 5.6 represent typical records obtained from tests of both high- and low-permeability zones and the effects of sand face plugging, choke sizes, and cushion weight. Again, reference to equipment and procedures, coupled with basic fluid-flow considerations, will serve to explain the records illustrated.

The test report provided by the service company will include either the metal pressure charts, film prints of the charts, or both. Also, a tabulation of pressure with time, as read from the charts, may be included. Otherwise the chart must be placed in a chart reader/scanner and such data obtained. In this case, a pressure-recorder calibration chart must also be available. It is an axiom that at least 10 pressure–time points should be obtained from each of the drawdown and build-up portions of the record. Commonly, the entire chart will be read at intervals of 5, 10, 15, or 30 minutes, depending upon the total time of each test period. Therefore, if such a routine is followed, the interval selected should be such that at least 10 pressure points are obtained from each drawdown and each build-up interval to be analyzed.

The computational analysis of drillstem test data is based on the equations for unsteady-state flow in a radial porous medium under flowing (drawdown) or shut-in (build-up) conditions. For the radial flow system, in the following pressure in the wellbore after a production time t can be expressed as:

$$P_{\mathrm{wf}} = P_{\mathrm{i}} - \frac{162.6\, q\mu B_{\mathrm{o}}}{Kh}\left[\log\left(\frac{Kt}{o\mu C r_{\mathrm{w}}^{2}}\right) - 3.23 + 0.87S\right], \tag{5.1}$$

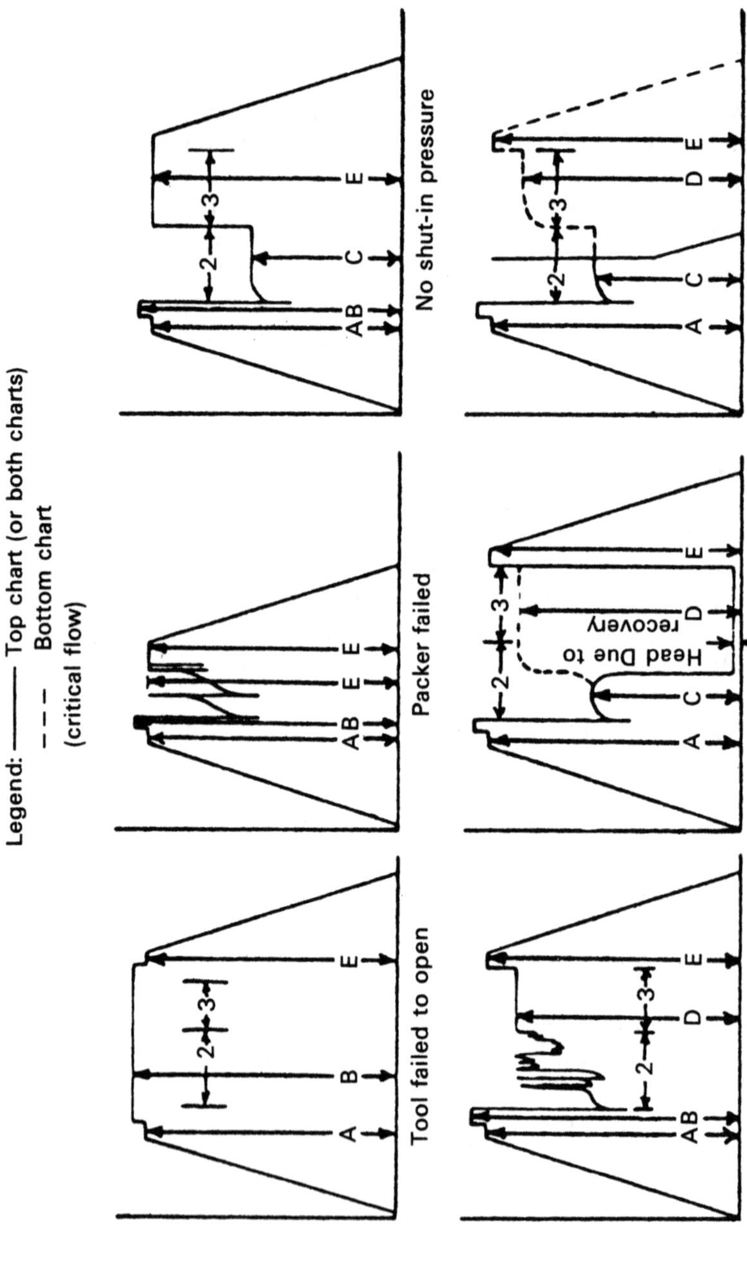

Figure 5.5 Mechanical failures during drillstem tests. (In Black, WM, A review of drillstem testing techniques and analysis, *JPT*, page 21, June 1956. Courtesy of SPE-AIME, © 1956).

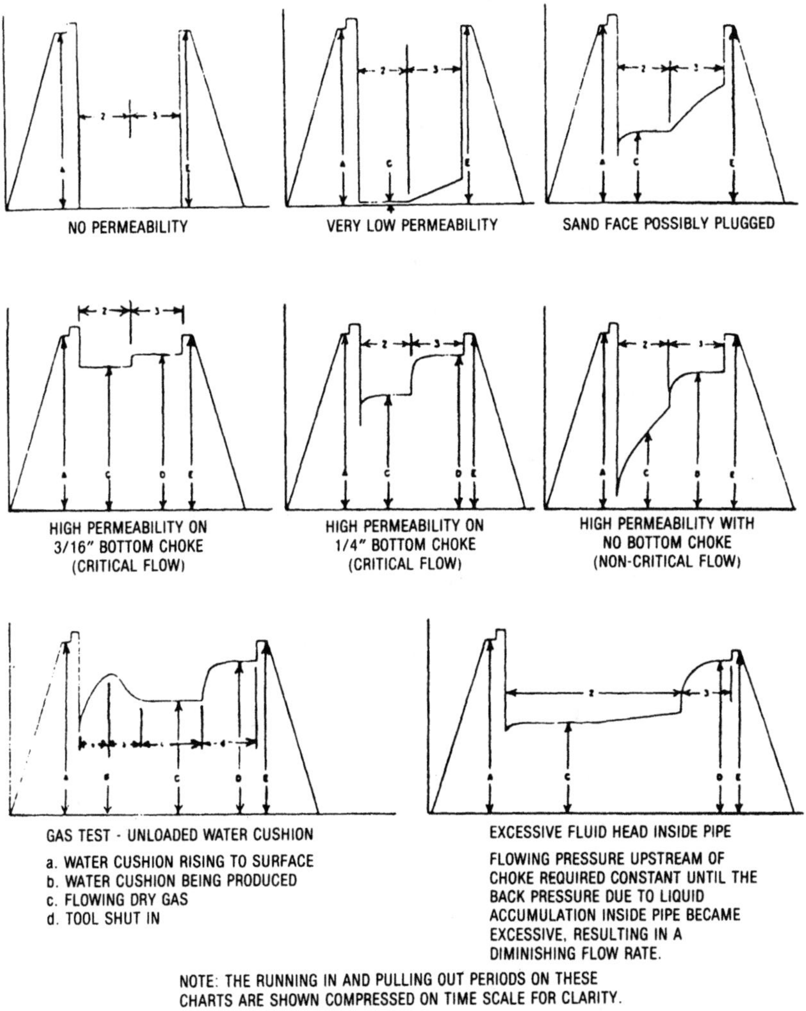

Figure 5.6 Formation and flow characteristics. (In Black, WM, A review of drillstem testing techniques and analysis, *JPT*, page 21, June 1956. Courtesy of SPE-AIME, © 1956).

where P_{wf} = bottomhole flowing pressure (psi)

P_i = bottomhole static pressure (psi)

q = oil flow rate (stock tank bbl/day)

μ = oil viscosity (cp)

B_o = oil formation volume factor (bbl/stock tank bbl)

K = effective permeability to oil (md)

h = formation thickness (ft)

t = production time (hours)

o = hydrocarbon porosity (dimensionless)

C = fluid compressibility (vol/vol/psi)

r_w = wellbore radius (ft)

S = skin factor for formation "damage" defined by:

$$S = \left(\frac{k}{K_s} - 1\right) \ln \frac{r_s}{r_w} \tag{5.2}$$

and

$$P_{skin} = S \frac{q\mu}{2kh} \tag{5.3}$$

These parameters are determined from the test data.

An examination of equation (5.1) indicates that during the transient-flow period of the test, a plot of bottomhole pressure versus log t should be linear, as shown in figure 5.7. Furthermore, the Kh product can be expressed as

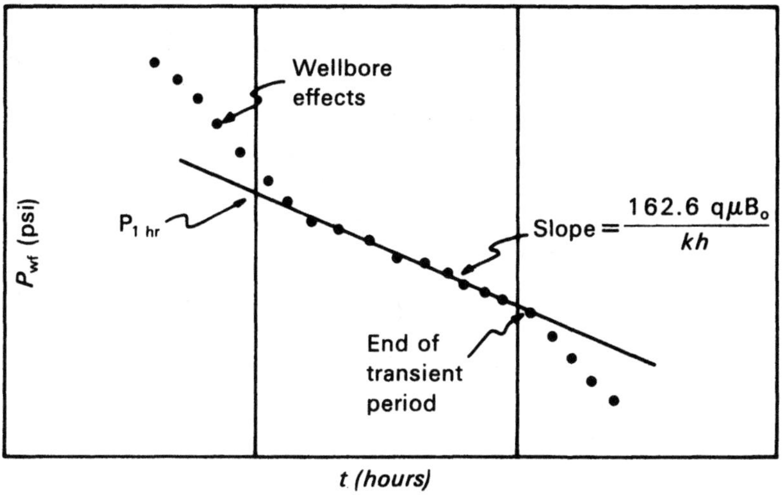

Figure 5.7 Transient drawdown analysis.

$$Kh = \frac{162.6\ q\mu B_o}{m}, \tag{5.4}$$

where M is the slope of the transient line in psi/$\log_{10}$cycle. Therefore, from the slope of the transient portion of the drawdown curve the formation product Kh can be obtained.

It can be shown further that the skin factor S can be expressed as:

$$S = 1.15 \left[\frac{P_1 - P_{1\,hr}}{m} - \log\left(\frac{K}{o\mu c r_w^2}\right) + 3.23 \right] \tag{5.5}$$

and the $P_{1\,hr}$ is also noted on figure 5.8.

Therefore, if the static formation pressure is known and the slope of the transient curve can be obtained, the drawdown data can be used to calculate the skin factor S. It should be noted from equation (5.2) that a positive skin factor would indicate reservoir damage and a negative skin factor would indicate an improvement in permeability near the wellbore.

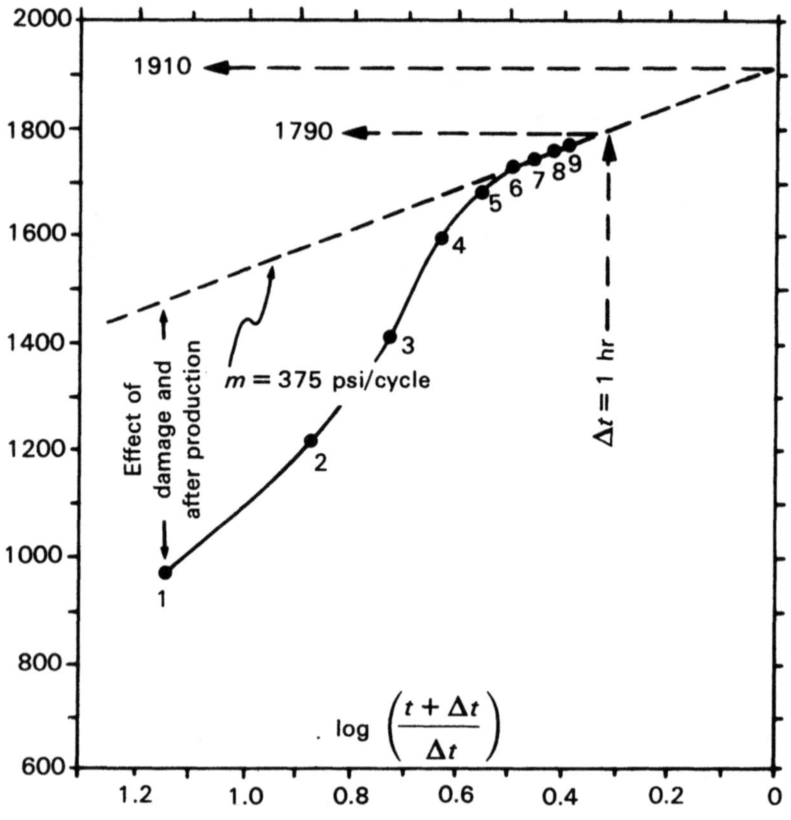

Figure 5.8 Pressure build-up analysis.

These same relationships can be used to analyze the pressure build-up portion of the test record. The pressure in the wellbore after the test tool is closed can be expressed as:

$$P_{ws} = P_i - 162.6 \left(\frac{q\mu B_o}{kh}\right) \log\left(\frac{t + \Delta t}{\Delta t}\right), \qquad (5.6)$$

where

P_{ws} = wellbore pressure (psi),

t = time the well was flowed (hours),

t = shut-in time (hours).

Likewise, the skin factor can now be expressed in a manner analogous to equation (5.6) as:

$$S = 1.151 \left[\frac{P_{1\,\mathrm{hr}} - P_{\mathrm{wf}}}{m} - \log\left(\frac{K}{o\mu c r_{\mathrm{w}}^2}\right) + 3.23 \right], \qquad (5.7)$$

where P_{wf} = the flowing pressure at shut-in time (psi).

This time the pressures are plotted versus the logarithm of the ratio $(t + \Delta t)/t$ and should yield a straight line after the period dominated by wellbore effects and after production, as shown in figure 5.8. The data plotted on this figure are given in table 5.1 as ready from a drillstem test pressure record. Here, time is in minutes, and the previous flow period lasted a total of 65 minutes (t).

As shown in figure 5.8, a simple extrapolation of the straight-line portion of the curve to a logarithm-scale value of zero, where $(t + \Delta t)/\Delta t = 1.0$, or infinite time, yields the static reservoir pressure of 1910 psi. Also noted on this plot is the pressure of approximately 1790 psi, which corresponds to the $P_{1\,\mathrm{hr}}$ and can be used with the final flowing pressure and the slope, m, to obtain the skin factor. Once again, the slope of the build-up curve can be used to obtain the formation product Kh.

TABLE 5.1 DRILLSTEM TEST PRESSURE BUILD-UP DATA

Point	Shut-in time, t	Pressure, P_{ws}	$\frac{t+\Delta t}{\Delta t}$	$\log\frac{t+\Delta t}{\Delta t}$
1	5	965	14.000	1.146
2	10	1215	7.500	0.875
3	15	1405	5.333	0.727
4	20	1590	4.250	0.628
5	25	1685	3.600	0.556
6	30	1725	3.167	0.500
7	35	1740	2.857	0.455
8	40	1753	2.625	0.419
9	45	1765	2.444	0.388

Example courtesy of Johnston-Macco.

Realizing that the pressure effects of a short-duration flow period, such as that of a drillstem test, only exist over some finite distance, we can determine the so-called radius of investigation:

$$r_i = \left(\frac{Kt}{40\ o\mu c} \right)^{1/2}, \qquad (5.8)$$

where in this case t is the flow time in days, and r_i is the radius of investigation in feet. The permeability is that determined from the build-up or drawdown test analysis.

This calculation, coupled with obvious reservoir depletion during the drillstem test, can also be used to ascertain the approximate size of a small, limited reservoir.

The presence and approximate distance to nearby flow barriers (such as faults or permeability pinchouts) and flow discontinuities (such as gas–oil or gas–water contacts) will be evidenced by abrupt changes in the slope m. Fractured reservoirs will also often behave in an anomalous fashion. Quantitative techniques are available for analysis of such conditions but they are beyond the scope of this course. For further methods of analyzing well test data, the student is referred to Monograph 1, *Pressure Buildup and Flow Tests in Wells*, SPE of AIME (1967).

5.5 Problems and Remedies

Most of the problems encountered in drillstem testing have a technical explanation and can be avoided entirely or at least made tolerable by proper planning and execution. The initial planning for a drillstem test should begin when the well is being planned and should be based on all that is known about the formations to be tested and the entire section that will be exposed in the open hole. Rather than being a one-man job, this planning should be done by personnel with all the various technical skills appropriate to the value, expense, and risks of the proposed test program. Likewise, the execution of the test should be a team effort between the drilling operators, the engineers, and geologists responsible for evaluating the well, the engineers who actually set up the drillstem test, and the personnel of the involved service company.

Outside of mechanical failure of the tools themselves, typical problems encountered include sloughing of formations, which can plug the anchor screens, or cause the pipe to stick, differential sticking of the pipe or the tools, pressures that exceed the capacity of the drillstring or surface equipment, packer leaks or packers failing to seat due to poor depth measurements or poor selection of a packer-setting point, out-of-gage hole that can cause the tools to stick or surging and swabbing effects, improper choke size selection, and the presence of noxious hydrogen sulfide gas without proper safety precautions having been taken.

Solving or avoiding these problems involves proper design and preparation when they are known to exist from past experience in the area. For wildcat wells, proceeding with caution and with careful preplanning is the only solution. This may not be sufficient to avoid certain problems, but dangerous conditions and remedial expenses can at least be minimized.

To test any zone, a proper and suitable packer setting point must be found. First, any formations used as a packer seat should be competent and able to withstand the forces imposed upon it. In any case, tight and impermeable sandstones, limestones, and dolomites are preferable to any type of chert or shale. Second, the formation must be thick enough for the packer assembly and must, at the same time, provide a reasonable margin for error in drillpipe tallies and wireline measurements. This usually means a minimum thickness of at least 15 ft, especially at considerable depths.

Sloughing formations and associated hole fill usually constitute the most serious potential problem. Drilling mud treatments or even mud change-outs may be necessary to minimize sloughing, but unless remedial steps are taken before sloughing starts, only minimum improvement can be expected. When the sloughing interval is above the packer, the stationary drillstem test equipment can be severely reduced, and it may even by necessary to run a string of casing before testing. When sloughing is below the packer depth, additional hole below the test interval may be necessary to accomodate hole fill and to decrease the hazard of plugging the anchor or choke. In such a case, a straddle test would probably be preferable since a conventional anchor could wash or suck its way down when the test tool is opened.

The hole should be to gage over its entire interval to prevent the problems previously mentioned and so that packer rubbers are not required to expand more than a reasonable minimum. (the smaller the expansion distance of a packer rubber, the higher the differential pressure it can stand.) Although little can be done about washouts outside of setting casing, undersized hole may require reaming.

Differential sticking can be a serious problem with stationary pipe. Take, for example, the situation when the pipe contacts the side of the hole at a permeable formation and the pressure in the hole is sufficiently higher than the formation pressure, so that the pipe is "frozen" in place. Rotating the pipe out of this situation requires that both the pressure forces and the mudcake friction be overcome. Avoidance of this problem is the best remedy. Drilling mud alteration is the usual preventative measure, including a decrease in mud weight, a decrease in mudcake thickness, and an increase in the "slickness" of the mudcake.

Obviously, the drillstring must be in good condition and it and all surface equipment and fittings must have sufficient pressure ratings and capacities to withstand the pressures and forces that may be encountered while testing. Pressure control to prevent burst pipe, surges, and other problems can be achieved by several means. Chokes can be used both in the bottomhole test assembly and at the surface. Water or inert-gas (nitrogen) cushions can be used in the drillpipe, and backpressure control valves can be installed. Every method, though, has its disadvantages and limitations. Chokes provide pressure control only under flowing conditions. Cushions are effective only while in place and are, therefore, only an initial control in many tests. Backpressure control valves overcome this objection but serve as an obstruction that can limit inside drillpipe operations.

In most wells, the drilling fluid that meets normal drilling requirements is satisfactory or can be adapted for drillstem testing. Treatments and changes range from simple filtrate-loss reduction and weight adjustment to the use of special muds, such as invert emulsions. As much as possible, such changes should be programmed during the planning stage so that the proper hole and well environment is already available at the time of testing and a minimum of corrective measures are necessary at the last moment.

The presence of hydrogen sulfide in the test fluids creates a special problem. This gas can be extremely toxic and, unless is can be contained and directed safely away from rig personnel and any nearby installations or habitations, it may be necessary to pass up the testing of such zones.

5.6 Drillstem Test Rules of Thumb

1. Generally, a mechanical-pressure measurement may be assumed to have leveled out or stabilized if it holds the same pressure deflection for 15 minutes or longer.
2. Shut-in pressure breakdowns should have a minimum of 10 increments.
3. Johnston 36-hour clock motors should show travel of 0.020 ± 0.005 inches/minute and the 72-hour clock motor should show travel of 0.010 ± 0.005 inches/minute.
4. A minimum of four points should be used to ascertain the presence of any straight line to determine steady state conditions.
5. In the absence of any other information, it is recommended that a minimum of 60 minutes be given to the initial shut-in period.
6. When the blow dies, shut the tool in for pressure buildup.
7. Good well: the flow to the surface is $FSI = \frac{1}{2} \times FT$, where FSI is formation shut-in and FT is flow time.
8. Average well: blow but no flow; $FSI = 1 \times FT$. Do not short cut.
9. Poor well: blow dies; $FSI = 2 \times FT$.

6
Offshore Rig Types

6.1 Platform Rig

The decision to use a platform rig should be made prior to the design of the platform. This decision would be influenced by size of field, depth of production, radius of drainage of platform, volume of production, availability of rigs and water depth.

The platform rig would consist of the basic rig components with the living accomodations being part of the platform. This would require additional space and strength to handle the loads imposed. The day cost of this type rig would be less than others, but could be offset by the additional expense of the platform. The rig would be, perhaps, a land rig that would be loaded on a barge and towed to the platform. It would then be hoisted on to the platform and positioned over the well template for drilling.

6.2 Jackup Rig

The decisions to use a jackup rig would be influenced by the same factors, but generally a lesser number of wells and shallow water would be strongly in favor of the jackup.

The platform design should allow sufficient room to position the jackup around the platform without damaging it. This would require the operator to look at a particular rig and sign a contract for this rig in order to assure the compatibility of the two.

The rig would be spotted near the platform with tugs and work boats. The rig anchor system would be put out and the tugs would assist in spotting the rig over the platform.

The rig anchor pattern should insure good holding power and not interfere with the normal operations of the platform such as pipelines, crew and work boats, etc. Anchor piles are sometimes installed on the platform to reduce the anchor-handling problem.

Once elevated, the jackup is the most stable of all drilling units, and drilling from a jackup is like drilling on land. The jackup is the least expensive type of rig to build and operate in water depths of more than 50 ft. The water-depth limit is about 350 ft.

Jackups have poor mobility because of their hull shape which makes them difficult to tow, and several tugs may be required to maintain a speed of 3 to 4 knots.

Jacking procedures are critical and the following requirements should be observed. The platform is assumed to be afloat with the spud cans completely filled. The operational instructions and preliminary check list preparatory to going-on-location have been properly executed:

1. Verify that good weather will prevail throughout this operation and that the platform roll and pitch do not exceed the limits described on the design curves for platform.
2. It is recommended that the platform engage the bottom and be elevated only during periods of daylight.
3. The ocean bottom should be clear of obstacles that may damage the spud cans.
4. Lower the legs in sequence to allow sufficient time for generator recovery before commencing the lowering of

the next leg. Stop each leg when it is about 10 to 15 feet above bottom.

5. Check wind, wave, and current direction and position the platform with tugs and anchors to minimize swing and/or drift. **WARNING:** If the platform is allowed to drift or swing as the spud cans contact the bottom, severe damage to the spud cans and/or to the legs may occur.
6. Lower the legs to the ocean bottom and continue to elevate the platform to the desired preload height, maintaining the platform level within 0.3 of one degree during elevating. Do not allow the distance between wave crest and the bottom of the hull to be less than 5 feet while elevated.
7. Preloading:
   a) If the variable load on board is less than the maximum allowable, determine the amount of load necessary to simulate the maximum allowable variable load and fill the preload tanks with this amount of seawater equally distributed on all legs.
   b) Simultaneously load all legs with the specified amount of preload providing an even loading on all legs. Maintain the platform level to within 0.3 of one degree by adjusting the rate of preload flow or, if necessary, by lowering the high corner of the platform. **DO NOT** attempt to elevate the platform with any preload other than that used to simulate the maximum allowable variable load on board.
8. Maintain reload until all settling has taken place and platform is stabilized.
9. Discharge all preload.
10. Elevate the platform to the desired drilling height maintaining the platform level within 0.3 of one degree at all times.
11. Upon completion of elevating, turn the console master key to the OFF position.

The jackup casualties fall into the following categories:

Undertow = 36%.

Moving on/off = 27%.

Blowouts = 27%.

Severe storms = 10%.

6.3 Semisubmersibles

The increased depth of operational waters has been met with the semisubmersibles. The development of subsea equipment has been such that the continued use of the semis is insured.

The semi (or floater, as it is sometimes referred to) is towed to the location buoy and spotted. The tugs will run the anchor pattern that has been previously selected for minimum interference with operations and weather. The semisubmersible will then be positioned over the proper location and maintain its position automatically within preset limits.

Some are self-propelled or have thrusters to assist in towing. The self-propelled speed is generally low, and some government regulations require tugs.

The semisubmersibles have very good motion characteristics that permit drilling operations to continue in waves of 35 to 40 ft high. Wind and current forces tend to push the unit off location. This calls for a heavy mooring system that uses chains (instead of wire rope) and 45,000 lb anchors. This system requires good handling.

Prudent operations are called for in force-8 weather with vertical rig motion greater than 5 ft and pitch or roll 2° either side of vertical with chain tension of 125 kip. The semisubmersible should be shut down in force-8 or forecast force-9 weather with vertical motion 8 ft and pitch or roll 4° either side of vertical with 150 kip chain tension.

The statistics of casualties on semisubmersibles are:

Severe storms = 66%.

Blowouts = 22%.

Undertow = 12%.

In the United States, the Coast Guard has been assigned the responsibility for developing measures to ensure the safety of life at sea. The American Bureau of Shipping (ABS) determines the maximum safe draft to which a drilling unit may be loaded and requires that a visible marking be placed on the vessel at this draft. This maximum draft, or load line, is indicated by the Plimsoll Mark. Samuel Plimsoll was a member of the British Parliament and was largely responsible for the passage of the British Merchant Shipping Act of 1876. This act called for the placing of a mark at the maximum safe draft in order that all persons on board the vessel could see that it was not overloaded.

Experience with vessels at sea continues to be evaluated to determine whether changes should be made in the laws to make vessels more seaworthy. In the case of drilling units, the load-line regulation applies to all units when they are floating. If a load line is assigned to a bottom-supported unit, it does not apply while the unit is being lowered or raised from the seafloor.

Drilling units have reserve buoyancy. This is defined as the total weight that would have to be added to cause it to sink.

On floating drilling units, there is some elevation above which the structure of the unit is not watertight. This is usually associated with a particular deck called the freeboard deck. The distance from the waterline measured vertically up to this deck is called the freeboard. The volume of the unit from the waterline up to the freeboard deck is called reserve buoyancy. Reserve buoyancy is the standby buoyance which is required in order to withstand the forces of the wind, waves, current, accidental flooding, or the shifting of weight aboard the unit.

Stability is the ability of a floating vessel to remain upright or return to the upright position when subjected to the forces of nature or operation. Therefore, certain rules must be followed by the operating crew if the unit is to perform safely.

The process of running casing or drillpipe affects the stability of a drilling unit, as does pulling out of the hole and racking the pipe in the derrick.

Drilling units are designed for adequate stability in the event of damage. No unit can survive extensive damage, but damage, if properly handled, need not be hazardous.

6.4 Drillship

Drillship is a ship with the center part hollowed out for drilling activity. This moonpool area is positioned by the ships power system and maintained within predetermined limits. Drillships are generally used for the remote wildcat type operations, or in very deep water, or for scientific exploration. The water depth is limited only by its mooring system.

Wind, wave, and current action tends to force the unit off its location. A mooring system of up to 10 anchors and mooring lines could be used to offset forces in any direction. The deployment and recovery of this system is difficult and hazardous. It cannot be accomplished safely in waves of more than 8 to 10 ft.

An alternative approach is dynamic positioning which requires a computer-controlled system. This system will sense the vessel's position relative to the well and direct the thrust of the propulsion units to maintain proper position. The power necessary to hold position by dynamics means quickly increases under high environmental loads, so that dynamically positioned vessels are not particularly suited for operations in areas of rough water.

The optimum operational water depth for the conventional (mooring lines and anchors) and dynamic systems overlap in the 1500-ft water depth range.

Drillships are the least expensive floating units to build by conventional mooring. The addition of dynamic positioning increases construction cost substantially, but no other alternative is presently available for ultradeep water.

The disadvantage to the drillship is its susceptibility to wave action. In 20 to 25 ft waves, the vessel may heave 8 to 10 ft. Drilling operations are usually suspended when heave reaches 5 to 7 ft, so in rough-water areas much of the time is spent waiting on the weather.

6.5 Submersible Rig

The submersible rig is similar to the jackup rig in that they both rest on the bottom. The submersible has hulls on which it floats while being towed to the location. Upon reaching the location, the hulls are flooded and sunk to rest on the bottom. The drilling deck is supported by columns from the hull and is well above the water level. Submersibles have a fixed operating water depth like the jackups and both are limited to drilling in relatively shallow waters. The submersible and jackup rigs provide a very stable platform for operations.

Upon reaching its location, the submersible is submerged by flooding one end of the hull to a reasonable tilt angle. The other end is then flooded, and the vessel is more or less rocked to bottom. The drilling deck must be at such an elevation that the waves can pass safely underneath. On leaving the location, the submerging process is reversed. (Some units submerge at a nearly level condition.)

Upon reaching the seafloor, the unit is leveled and all tanks are flooded to bring the unit to the desired bearing load on the bottom.

Care must be taken to ballast or deballast the unit evenly about the center line. Too much weight on one side can bring the unit to the surface heeling to one side. Planning can eliminate this problem. A careful operator knows the weight of the unit and how much ballast must be removed to cause it to float freely off bottom. As the unit starts to break suction from the seafloor, observant rig personnel can detect the first movement of the unit as it starts to move with the waves. The deballasting operation should be stopped; the force of the sea gradually causes water to permeate the soil beneath the hull and allows the unit to float. The unit is then raised to the surface.

Forcing the unit free from the seafloor can cause dramatic movement of the unit which could cause loose equipment on deck to shift and the unit to come free with a severe heel or trim.

Casualties on submersible rigs are caused by the following:

Storm = 33%.

Blowout = 33%.

Undertow = 17%.

Moving on/off = 17%.

Notice that on submersible rigs 33% casualties are due to storms, while on jackup and semisubmersible rigs this factor is responsible for 10% and 66% of all casualties, respectively.

6.6 Offshore Rig Design Rules of Thumb

1. Potable-water requirements	65 gal per person per day
2. Living quarters	50 sq ft per person
3. Engine–generator deck area	10 kW/sq ft
4. Motor deck area	60 hp/sq ft
5. Engine deck area	16 hp/sq ft
6. Engine–generator deck load	260 lb/sq ft
7. Motor deck load	440 lb/sq ft
8. Engine deck load	170 lb/sq ft
9. Engine–generator weight	29 lb/kW

7

Offshore Environment

7.1 Transportation

Offshore transportation is accomplished by means of vessels or helicopters. The vessels are generally fast-moving aluminum-hull crew boats, but sometimes a slower workboat will be used.

The following safety precautions to protect all personnel from dock to rig deck are mandatory:

a) Engines should be started prior to any personnel boarding. This will help minimize chances of a disaster in case of an explosion.
b) No personnel should board a boat from the dock that is not securely tied to the dock.
c) Sign the passenger list as soon as possible after boarding.
d) Before leaving the dock, the Skipper must inform all passengers of the life-saving and fire-fighting equipment located on the boat.
e) Passengers should never ride on the outside of boats.
f) Passengers should not walk around but stay in their seats as much as possible.

g) Passengers should not be allowed in the pilot's cabin.
h) All passengers must put on life jackets before leaving the boat.
i) Care should be exercised in handling baggage during boarding the boat or the rig.
j) Both hands should be free when boarding rig.
k) Passengers should not be rushed to board the drilling unit and should leave boat on the rise.
l) If swinging ropes are used, they should be one inch in diameter with a knot every foot, so that personnel can hold on to them.
m) Ropes should be hung so that boarding personnel does not hit the structure.
n) Always make swing when boat is at its highest peak.
o) If personnel baskets are used, they should be equipped with stabilizers to keep the net in the upright position.
p) Hooks used on baskets should have safety latches.
q) Extra personnel should be on hand to assist passengers.
r) All personnel riding the basket should stand on the outside of the basket holding onto the net and wearing life jackets.
s) Return life jacket to basket upon reaching rig or boat.
t) If a helicopter is used, it must be understood by all personnel that the pilot is in charge and is the one who makes the decision to fly or not to fly.
u) Passengers must observe the no-smoking regulations.
v) Personnel must put on life vests, and fasten seat belts as soon as they enter the cabin.
w) Do not approach or leave the helicopter if the blades are still turning except upon a signal from the pilot.
x) Do not raise your head over top of heliport until helicopter has landed and reduced engine speed.
y) Never store luggage or cargo on heliport.
z) Furnish pilot with correct weights and always walk to the front of the helicopter—**NEVER** go around the tail. Should you be injured by the moving blades, the helicopter cannot fly you into port as it cannot be moved until inspected and repaired.

The landing pad for helicopters is a very important consideration for every rig; some pads are designed for specific helicopters. Knowing the pad design is very important for normal operations, but even more so in case of emergency.

7.2 Logistics

Supplying the offshore drilling unit with personnel, food, water, mud, drilling tools, pipe, fittings, and outside services requires a lot of planning as well as personnel:

a) Personnel have duty hours as well as days off. The working day is generally 12 hours long, but the schedule varies 1 week on—1 week off to 30 days on—30 days off.
b) The given number of personnel on the rig depends on the drilling operation. The range of accommodations is as follows:

	Small	Large
Jackup	52	75
Semisubmersible	62	84
Drillship	53	85

c) Needless to say, the health, safety and welfare of all personnel is essential from a logistics viewpoint.

7.3 Food

The food supply should be plentiful and good. People will work and not complain in a bad working environment if the food and beds are good. But if the food and beds are not good, then personnel will quit. This causes more logistic problems.

The quantity of food supplies depends on the environment and on the distance to the supply point, but with 75 to 85 mouths on board, with four meals being prepared daily, it could run into

TABLE 7.1 SUPPLIES FOR A LARGE VESSEL

	Quantity	Tons
Fuel	4,250 bbl	707
Potable water	825 bbl	148
Brake cooling water	460 bbl	85
Lube and gear oil		5
Drilling water	4,000 bbl	700
Wash water	1,000 bbl	180
Bulk mud and cement	7,060 cu ft	400
Liquid mud	1,700 bbl	600
Chemicals	6,000 sacks	300
Drillpipe, collars, casing, and riser		500
Chain, wire rope, and anchors		50
Misc. tools, equipment, and dry stores		40

several tons. A summary of supplies for a large vessel is given in table 7.1.

These quantities must be maintained, thus the supervisor will not let a vessel go to port with extra fuel or potable water because the rig can always use these items.

The drilling engineer/supervisor should not let unnecessary equipment stack up on the rig, because the load and the available space are very limited. When the surface casing is set, all bits, tool, and supplies for this operation should be secured and sent to shore if possible. Then the vacated space can be used for tools and equipment of upcoming operations. The drilling engineer/supervisor must work closely with the warehouse personnel to insure a steady flow of *proper* materials and supplies, because improper materials and supplies can create logistic problems and result in considerable expense.

Operations that could be hazardous should not be started until all equipment, supplies, and materials are on the vessel. The radio message "it's on the way" does not account for sudden storms, work/supply vessel breakdowns, or wrong materials sent out.

Special attention should be paid to the outside-services group to insure that their equipment, material/supplies are available and in good working order before they are needed. This would include loggers, cementers, divers, etc.

Near the end of drilling operations, with a move coming up, proper scheduling of workboats, tugs, supply vessels, helicopters and crew boats becomes a real juggling act.

7.4 Weather

Severe weather and sea conditions further complicate offshore drilling operations. The operational time in mild weather could be as follows:

Operation	*Percent*
Drilling and tripping	57.0
Fishing, logging/cementing	21.0
Running casing	7.0
Running BOP stack or riser	15.0
	100.0

Severe sea and weather conditions could change this to:

Operation	*Percent*
Drilling	15.4
Tripping	11.8
Running casing & cementing	11.0
Circulating	4.9
Reaming	2.1
Coring	1.4
Logging	5.5
Drillstem test	2.8
Plug & abandon	2.8
	57.7

The remainder of the time would be distributed in the following way:

Waiting on weather	20.8
Mooring & anchoring	8.3
Running & retrieving stack	8.3
Other	4.9
	42.3

The wind speed is generally measured on the Beaufort scale (table 7.2). The wind is said to "blow into the compass," e.g., a northwest wind blows from the northwest. The shaft of a wind arrow used on weather maps indicates the direction and the feathers show the speed of the wind.

The offshore personnel are very concerned about the weather report and most operators have more than one source of information. The first concern is the wind, but high winds create high waves (table 7.3).

TABLE 7.2 THE BEAUFORT SCALE FOR WINDS

Beaufort number	Description	Wind speed at 33′ elevations (miles/hour)
0	Calm	0
1	Light air	1–3
2	Light breeze	4–7
3	Gentle breeze	8–12
4	Moderate breeze	13–18
5	Fresh breeze	19–24
6	Strong breeze	25–31
7	Near gale	32–38
8	Gale	39–46
9	Strong gale	47–54
10	Storm	55–63
11	Violent storm	64–73
12	Hurricane	74–83

TABLE 7.3 THE BEAUFORT SCALE FOR WAVES

Beaufort number	State of sea (deep water)	Average significant wave height, ft
0	Calm and mirrorlike	0
1	Small ripples, no crests, no foam	¼
2	Small choppy waves	½
3	Large choppy waves	2
4	Small fully developed waves	4
5	Moderate waves with foam crests and possibly some spray	6
6	Large waves, most have foam crests	10
7	Waves increase in height, foam is blown in direction of wind	14
8	Foam blown in streaks, waves increase in height	18
9	Very high waves, wave crests begin to tumble & roll. Dense streaks of foam	23
10	Extremely high waves. Sea surface is white in appearance. Heavy tumbling in waves. Large patches of foam are blown	29
11	Sea covered by blowing foam. Visibility is reduced. Small ships may be obscured by high waves	37
12	Sea completely white with blowing foam and spray, so that visibility is very poor	45

The term *significant wave height* has been introduced by visual sea observers. It is defined as the average height of the highest one-third of waves in a sea state.

A sample of a weather forecast and the actual conditions on a semisubmersible are given in table 7.4.

Force-10 weather could require taking on ballast of 80 to 85 ft draft. At some point each rig will have a shut-down procedure to secure the equipment and personnel.

In certain parts of the world, work is performed during the weather window, which is a time interval during the year when the weather is favorable to the operator. This is also true on land, especially in the mountains. Vast networks of equipment and personnel for observing, reporting, and interpreting weather phenomena are on duty every hour of the day and night in numerous areas all over the world. Forecasts and warnings of

TABLE 7.4 WEATHER FORECAST & ACTUAL CONDITIONS

	Weather forecast	Actual
Force	9	9
Wind, mph	46–53	54–62
Pitch or roll	4–5°	5–6°
Heave, ft	10–15	15–20
Anchor tension, kip	125	150
Force	10	10
Wind, mph	60–70	63–71
Pitch or roll	4–6	6°+
Heave, ft	8–10	10
Anchor tension, kip	125	150±

hazardous conditions are routine. All of these services are available to anyone with adequate communications equipment.

An understanding of the hazards in a drilling location should be the first order of business where planning of equipment specifications, supplies, logistics, and personnel is required for a successful operation. After operations begin, a close watch should be kept through weather reports and forecasts to prevent unexpected situations which can be very costly.

7.5 Communications

Mobile drilling units are equipped with many different types of communications systems, some of which are VHF/FM radio, HF/SSB radio, microwave radio, telex, telephone. Radio communications are controlled closely by governmental regulatory agencies. They generally require a licensed operator aboard the drilling unit.

Some regulations do not allow drilling units to have the 2182 kHz frequency which is used internationally as the emergency channel, however any unit classified as a vessel will have it. Other channels may be designated in an area for emergency operations. This could be the international radiotelegraph distress frequency 500 kHz.

The order of priority for communications in mobile service shall be as follows:

1. distress calls, messages, or traffic preceded by the international distress call "May Day"
2. communications preceded by the urgency signal
3. communications preceded by the safety signal
4. communications relating to radio direction/location
5. communications relating to the navigation and safe movement of aircraft
6. communications relating to the navigation, movements, and needs of ships and weather observations messages

It is the duty of radio operators to report to their supervisors on any infringements of the radio regulations which they may detect. The drilling unit will keep a log of transmissions which may be viewed by authorities. In some areas, this log may be required as a report to the government agency on a scheduled basis.

For safety reasons, all radios are generally turned off during perforation operations. This action will be passed on to all interested parties prior to turning them off, and the announcement of being back on the air will come as soon as the operation is safe.

Radio language is often confusing to many, so an international letter system has been developed (table 7.5).

In some areas, several companies can be assigned the same frequency, so the radio traffic can be quite heavy for a particular unit. This is an inconvenience to all, and let us hope that it will not last very long. It is important to keep your messages brief and to the point. Many forms have been developed for this reason, such as the one shown in figure 7.1.

This form could be modified, but it must cover report data for engineer/geologist, mud engineer, barge engineer, and management.

It should be stressed that communications is the heart of the operations: when communications stop, the operation soon dies. Treat your communications system as if it were your heart—

TABLE 7.5 INTERNATIONAL LETTER SYSTEM

Letter	Word Used	Pronounced as
A	ALFA	*AL* FAH
B	BRAVO	*BRAH* VOH
C	CHARLIE	*CHAR* LEE OR *SHAR* LEE
D	DELTA	*DEL* TAH
E	ECHO	*ECK* OH
F	FOXTROT	*FOKS* TROT
G	GOLF	GOLF
H	HOTEL	*HOH* TELL
I	INDIA	IN DEE AH
J	JULIETT	*JEW* LEE *ETT*
K	KILO	*KEY* LOH
L	LIMA	*LEE* MAH
M	MIKE	MIKE
N	NOVEMBER	NO *VEM* BER
O	OSCAR	*OSS* CAH
P	PAPA	PAH *PAH*
Q	QUEBEC	KEH *BECK*
R	ROMEO	*ROW* ME OH
S	SIERRA	SEE *AIR* RAH
T	TANGO	*TANG* O
U	UNIFORM	*YOU* NEE FORM
V	VICTOR	*VIK* TAH
W	WHISKEY	*WISS* KEY
X	X-RAY	*EKS* RAY
Y	YANKEE	*YAN* KEY
Z	ZULU	*ZOO* LOO

provide maintenance, do not let personnel abuse the radiowaves, keep proper records, simplify report forms for your operations, and emphasize these points to your personnel.

7.6 Planning the Well

Let us drill a 13,600 ft offshore well in the area of Sicily, where we will not be hit too hard with sea and weather conditions. The location could be latitude 36°44′22″ North and longitude 14°28′40″ East.

Since this is a wildcat or semiwildcat operation, we discussed

OPERATOR__________ WELL NAME______REPORT NO.______

WATER DEPTH__________ OPERATIONS/REPORT TIME__________

OPERATOR'S SUPERVISOR______ __________

TOOL PUSHER__________ __________

DEPTH TODAY__________ BOTTOMHOLE ASSEMBLY__________

PREVIOUS DEPTH__________ __________

FOOTAGE MADE__________ __________

HRS. DRILLED__________ __________

ROTARY rpm__________ WT. on BIT_____ PUMP PRESS______

BIT NO. __________ MUD RECORD WEATHER & BARGE DATA

SIZE __________ WT__________ 0600 Report Time of

MAXIMUMS

MFG. __________ Visc. sec_____ WEATHER__________

TYPE__________ WL-CC_______ BAROMETER_______

JETS 1/32_______ F.C. _______ WIND SPEED_______

DEPTH OUT_______ pH_________ DIRECTION_______

DEPTH IN_______ Solids %_______ SWELL (HT & DIR)____

TOTAL FIG._______ OIL %_______ SWELL PERIOD_______

TOTAL HRS RUN______ SAND_______ SEA (HT & DIR)_____

CONDITION_______ CHL_______ COMB. SWELL &______

DEVIATION DEPTH DEV DIRECTION SEA(HT&DIR)______

RECORD _____ ___ _______ PERIOD__________

_____ ___ _______ ROLL__________

_____ ___ _______ PITCH__________

_____ ___ _______ HEAVE__________

DP AND DC ON RIG ANCHOR TENSIONS (1000 pounds)

No. jts._____ SIZE_____ DP 1_____2_____3_____4_____5_____

No. jts._____ SIZE_____ DC 6_____7_____8_____9_____10_____

No. jts._____ SIZE____ DC 11_____12_____

PERSONNEL ON BOARD HOLE POSITION_______ RISER ANGLE_______

Figure 7.1 *(see caption opposite)*

CONTRACTOR________	VARIABLE DECK LOAD_____TONS DISPL______
OPERATOR________	DRAFT________
CATERER________	RISER TENSIONER PULL
DIVERS________	1______ 2______ 3______ 4______
CEMENTING________	SAFETY DRILLS
LOGGING________	TYPE______TIME & REMARKS______
VISITORS________	FIRE________
________	LIFE BOAT________
TOTAL________	BOP________
OPERATIONS-REPORT	PIT________
ACCIDENTS	HANG OFF________
________	OTHER________

Figure 7.1 Sample radio message form.

the well with the geological staff and agreed on the following definitions of the lithology:

1. Soft = soft shales, clays, red beds, soft limestones, weak sandstones, salt, gypsum and anhydrite.
2. Medium = hard limestones, sandstones, dolomite, and hard shales.
3. Hard = chert, dolomite, and sandy shales.
4. Extremely hard = quartzite, quartzitic sand, basalt, and chert.

We are going to drill a 26-inch hole. The first 2000 ft of the formation are expected to be soft. The lithology of the 2000 to 5400 ft interval is expected to be soft to medium. The 5400 to 10,150 ft area will be soft to medium, with insert bit probably drilling better in the medium formations. We still expect the formations at 10,150 to T.D. to be of the same hardness, but of better compaction; thus we go to the insert bit throughout this interval.

Additional discussions change the hole size depths, but this does not affect the lithology, only the drilling program.

7.7 Drilling Program

1. Buoy the location by using SAT-NAV.
2. Do seabed survey of location area: soil tests, sparker, side-scan sonar survey with divers' visual inspection.
3. Move rig to location, spot and preload to contractor's specifications.
4. Do full inspection of equipment to insure that it is all operable.
5. Make complete inventory of all materials.
6. Report any discrepancies in items (4) and (5) to rig manager and to operations manager.
7. Inspect blowout preventer by performing functional and pressure test as per manufacturer's manual.
8. Prepare the initial spud and drilling assembly.
9. Prepare the drilling fluids for spud.
10. Drill 36″ hole and set 30″ casing at 370 ft (114 m). See table 7.6.
11. Drill 26″ hole to 1620 ft (500 m) using up to 10.5 lbs/gal mud and viscosity as required; survey every 200 ft (66 m) and maintaining deviation below 1° at 1620 ft (500 m).
12. Drill 32 ft (10 m) of overhole below proposed casing-landing depth. Pull out of hole, pick up hole opener and run back to bottom. Circulate hole clean, pull out of hole, and run 20″ casing.
13. Run 1620 ft (500 m) plus or minus of 2″ H-40 94 lbs/ft casing. Run with float shoe and one centralizer on three bottom joints and one at 100 ft plus or minus below mud line; thread lock and strap-weld two bottom joints.
14. Cement as per cementing program: check for flowback, release drillpipe running string, backwash and pull out of hole.

15. Pretest blowout preventer stack, recheck equipment, materials and supplies. Nipple up 20″ BOP. Drill out in 12 hours.
16. Drill out shoe with 17½″ bit by using mud as per mud program. Drill 32 ft (10 m) below shoe. Pick up inside shoe. Test formation to the equivalent of 12.5 lb mud or leak off. Drill to 7776 ft (2400 m) by using bottomhole assembly depicted in figure 7.2. Survey 200 ft (66 m) below 20″ shoe and every 500 ft (155 m) thereafter or as requested by company representative. Maintain deviation of less than 3° at 7776 ft. Drill 32 ft (10 m) plus or minus over hole.
17. Circulate and condition hole; pull out of hole. Pick up hole opener and run to bottom for logging (if required). After a clean-wiper trip, pull out of hole and run electric logs as per logging program.
18. After log run, run in hole and condition hole; pull out of hole and run 13⅜″ casing string of P-110 68# and N-80. (See also the alternative casing program.)
19. Run and land 7776 ft. plus or minus of 13⅜″ casing; drift all casing to 12.250 inches. Run casing with float shoe and float collar on top of bottom joint. Centralize the six bottom couplings. Thread-lock and strap-weld two bottom joints. Centralize into 20″ with two centralizers.
20. Cement full length as per program. Confirm floats holding. Release running tool, backwash 20 in. BOP stack. Flush BOP stack through choke and kill lines.
21. Reseat tool and nipple down 20″ BOP. Nipple up 13⅜″ BOP and test as per manual through choke and kill lines.
22. Run cement bond log if required.
23. Drill out shoe and drill 32 ft plus or minus (10 m). Pull back into casing and run leakoff test. Drill ahead with 12¼″ bit by using mud program and bottomhole assembly, as outlined. Survey every 155 m to depth of 11,362 ft. plus or minus (3507 m) and maintain deviation of less than 4° at 11,362 ft.
24. Circulate hole clean if possible and log as directed.

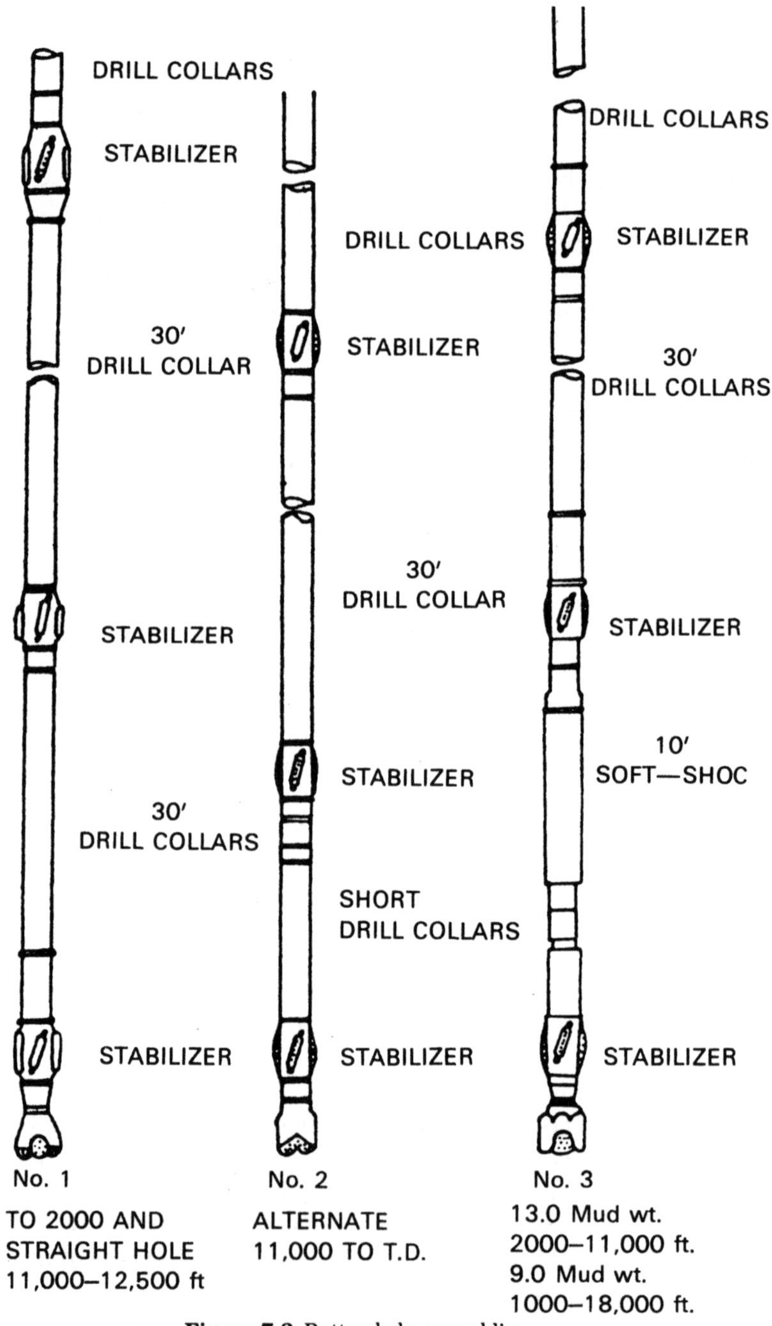

Figure 7.2 Bottomhole assemblies.

25. Run and land 9⅝" casing at 11,362 ft. ±. Drift all casing to 8½ inches. Run casing with float shoe and float collar on bottom joint with centralizers on three bottom couplings. Threadlock and strap-weld two bottom joints. A multiple stage cementer collar will be used, the multiple stage cementer collar depth to be determined after logging.
26. Cement as per program.
27. Nipple down BOP. Install 13⅝ × 10 spool.
28. Nipple up 13⅝ inch 10,000 # BOP stack. Pressure test as per prognosis.
29. Run cement bond log and other logs as required.
30. Drill out shoe and drill ahead 32 ft plus or minus (10 m) with 8½" bit. Run leakoff test. Drill ahead with 8½" bit and bottomhole assembly as per program. Survey every 155 m or as requested and maintain deviation of less than 6° at total depth of 13,608 ft. plus or minus (4200 m) by using mud program as outlined.
31. Evaluate at total depth as per evaluation program—condition hole.
32. Run 7" liner total depth to 200 ft inside 9⅝" casing by using liner hanger with tieback attachment. Cement as per program.

7.8 Drilling-Fluids Program

The following program is based on available data and represents the best judgment of those involved in its preparation.

The on-site mud engineer will monitor the mud properties continuously and, if warranted, will recommend changes to the drilling supervisor.

Depth interval(ft)	Mud wt., (ppg)	Visc. (sec.)	Water loss (ml)	System
0–2000	8.5–8.8	60	no control	Seawater
2000–5400	8.8–9.5	60	20	Seawater
5400–10,150	8.8–10.2	50	6	Seawater
10,150–13,608	9.5–11.5	40	4	Seawater

TABLE 7.6 BIT AND HYDRAULIC PROGRAM

Gross interval footage	Hole size	Bit type	Footage Bit (depth out)	Bit wt. (min.)	Rotary Speed min. max.	Vel. ann.	Pumps no.— spm	gpm	Nozzles	Bottomhole assembly
0–2000	26	Hughes OSC 3A	2000	5–20	120–150	45	2–60	850 plus	Conv.	9″ drillcollars + stab at 60 ft
2000–5400	17½	Security S35-J M4NJ-S84 Smith DSJ	5400	30–80	80–150	105	2–65	820 plus	3–20 3–24	Soft shoe and assembly No. 3
5400–10,150	12¼	Smith *F–3SDT Security M44N–*S33	10,150	40–80	80–120 *50–60	125	1–65	650 plus	3–15 3–18 3–20	Soft shoe and assembly No. 3
10,150–T.D.	8½	Security *S84F- *S86F *S88F M44N M77 *Smith F–2, *F–3, F–4	11,860	30–60	*50–60 60–80	150	1–33	320 plus	3–10 3–12 3–16 3–18	Soft shoe and assemblies No. 3 and 2

*All journal bearing bits will have a maximum rotary velocity of 60 rpm.

NOTE:

a) For all logging raise viscosity to 60.
b) You may have to add salt to seawater off coast of Sicily to raise its weight to 10.2 ppg; add soda ash to raise weight to 10.6; add sea mud or drispac to carry weight higher; add starch to control water loss.
c) This system would require seawater to be filtered.
d) Loss of circulation is possible, so all solids-control equipment will be operating at maximum efficiency to maintain a minimum mud weight.
e) Bentonite (if used) should be prehydrated. Mud weight, viscosity, and water losses should be adjusted to maintain maximum penetration rate consistent with a clean, stable wellbore.
f) If gumbo clays are encountered, a drilling detergent may be added to minimize balling. If sloughing shales become a problem, mud weight may be raised and additives used to minimize the problem.
g) The use of seawater (brine) should minimize the swelling of clays thus maintaining a good wellbore.
h) Constantly monitor the pits to detect loss or gain.

7.9 Casing and Cementing Program

1. 30″ pipe already set.
2. 20″ pipe 0–1620′ ± (500 meters)
    a) Run float shoe on bottom with float collar one joint up. Run one centralizer on each three bottom joints and one inside 30 inch. Thread-lock and strap-weld two bottom connections.
    b) Keep casing full while running in hole.
    c) To minimize pressure surge, do not run in hole faster than 3 ft/sec.
    d) Land casing in 30-inch well head.

   e) Run in hole with drillpipe: break circulation and circulate hole volume, then cement through the drillpipe; check for flow back in 30-inch annulus. Bump plug with 500 psig, hold for 5 min, and release.

3. 13-3/8″ casing 0–7776 ft. ± (2400 meters)
   a) Run float shoe on bottom with float collar one joint up.
   b) Run one centralizer on each of the six bottom couplings and two centralizers inside the 20 inch casing.
   c) Thread-lock and strap-weld the two bottom joints.
   d) Keep casing full while running in the hole.
   e) Do not run faster than 2 ft/sec.
   f) Land casing and circulate bottoms up or until clean returns.
   g) Cement and displace, bump plug with 900 psig, hold 5 minutes and release.
4. 9-5/8 inch 0-11, 364 ft. ±
   a) Run float shoe on bottom with a float collar one joint up. Run multiple stage cementer tool at 7900 ft plus or minus with a metal pedal basket and one centralizer below and one centralizer above multiple stage cementer tool.
   b) Run one centralizer on each of three bottom joints and two centralizers inside 13⅜″ casing.
   c) Should lost circulation zone be encountered in this interval, consideration should be given to the cleavage barrier or postplug method of cementing arrangement.
   d) Thread-lock and strap-weld two bottom joints.
   e) Keep casing full while running in the hole; do not run faster than 2 ft/sec.
   f) Land casing and circulate bottoms up with *caution*; circulate until clean returns appear prior to cementing.
   g) Cement and bump plug on first stage with 1000 psig; hold for 5 min and release. Open multiple

stage cementer tool and circulate hole for 4 hours, then cement second stage. Each plug on second stage at 1000 psig. Hold for 5 min and release.

5. 7 inch liner 11,162 ft. to T.D. 13,608 ft. ±.
   a) Run float shoe on bottom with float collar one joint up.
   b) Run centralizer on each of three bottom joints and one inside 9⅝ casing.
   c) Thread-lock and strap-weld two bottom joints.
   d) Run liner hanger assembly with tie-back attachments and set at 11,162 ft ±. Do not run faster than 2 ft/sec.
   e) Circulate bottoms up or until returns are clean.
   f) Cement liner, bump plug with 1000 psig, hold 5 min and release.
   g) Clean off top of liner hanger assembly.
6. General procedures for cementing:
   a) Preflush ahead of all cement jobs.
   b) Reciprocate pipe enough to insure good movement at total depth.
   c) Displace at equivalent of 90 to 120 ft/min annular velocity until tail in slurry reaches shoe, then slow to 2 to 4 bbls/min while displacing tail-in around shoe.
   d) Monitor all mud and cement volumes closely. Gauge tanks prior to, during, and after cementing.
   e) Keep all mixing and displacing pressure recorders. Note sudden changes of pumping rate or other conditions affecting job.
   f) Continuously sample slurry density while mixing. Take at least four samples of each slurry mixed and observe the setting time in a heated atmosphere.
   g) Measure displacement accurately by using tank measurements and stroke counters.
   h) Use caliper logs if possible to determine proper volumes.

i) Use programmed excess cement for 20-inch casing.
j) Use caliper plus 40% for 13⅜" and 30% for 9⅝" and 7" casing.
k) All cement slurries will be checked with a compressive strength tester.
l) Run cement bond log prior to drilling ahead if practical.

7.10 Calculations for Cementing

1. 20" pipe depth 1620 ft, hole size 26 inch. Use class A cement mixed with seawater. Note: If feasible, add 2% calcium chloride to last 50 bbl of mixing water.

Open Hole Annular Volume = 1220′ × 1.505	= 1836
Excess 100%	= 1836
30" annular volume 400 × 2.405	= 962
	4634 ft

Slurry yield	= 1.16 ft/sack, 15.9 lbs/gal
Require	= 3995 sacks class A
	19,969 gallons seawater

2. 13⅜"; casing depth 7776 ft, hole size 17½" Tail-in with 300 sacks class G Neat. Tail-in slurry: 300 sacks plus seawater (1500 gallons)

Slurry yield	= 1.16 ft/sack
Slurry weight	= 15.9 lbs/gal
300 sacks	= 348 cu ft
Open hole volume	= 3401 cu ft
Excess volume 40%	= 1360 cu ft
Annular vol. 20"	= 1651 cu ft
	6412 cu ft

Lead slurry volume 6412 − 348 = 6064 cu ft. Class G

cement + 2% prehydrated bentonite in seawater.

Slurry yield	= 1.6 cu ft/sack
Slurry weight	= 14.0 lb/gal
Materials required	= 3790 sacks class G, 10,456 gal fresh water, 20,910 gal seawater, 6999 lb Bentonite.

Bentonite should be prehydrated in fresh water. The volume can then be made up to 25,923 gal with seawater with no adverse effect. Total requirements for 13⅜" casing:

4090 sacks class G cement
10,456 gallons fresh water
22,410 gallons seawater
6999 lb Bentonite

3. 9⅝" depth 11,326 ft. plus or minus, hole size 12¼" Multiple stage cementer tool set at 7900 ft plus or minus.
   a) First stage 11,326 to 7900 ft plus or minus
   b) Tail-in 300 sacks of neat salt-saturated class C

cement—seawater 1398 gal, salt	= 3868 lbs
O.H. volume 3426 × 0.3132	= 1073 cu ft
Allow 30% excess	= 322 cu ft
	= 1395 cu ft

Tail-in slurry with 1.23 cu ft/sack yield = 369 cu ft. Lead slurry volume 1395 − 369 = 1026 cu ft. Lead slurry is prepared from class G plus 2% prehydrated Attapulgite in saturated saltwater

Slurry weight	= 14.26 lbs/gal
Slurry yield	= 1.84 cu ft/sack
Seawater	= 9.11 gal/sack
Salt addition	= 2.77 lbs/gal (116 lbs/bbl)
Attapulgite	= 0.167 lbs/gal (7 lbs/bbl)

For 1026 cu ft slurry requirements are:
558 sacks class G cement
5084 gal seawater

849 lb Attapulgite
14,057 lb salt

c) Second stage 7900 to 0 ft.
O.H. Annular volume = 100′
± × 0.3131 = 31 cu ft
30% excess = 9 cu ft
40 cu ft

7800 to 0 in 13⅜″
7800 × 0.3354 = 2616 cu ft
2656 cu ft.

Tail-in with 100 sacks of Neat
Salt saturated class G slurry
Seawater = 1367 gal
Salt = 3787 lb
Tail-in slurry yield 1.25 cu ft/sack
Slurry volume tail-in = 125 cu ft
Lead slurry prepared from class G plus 2% prehydrated Attapulgite in saturated salt water.

Slurry weight = 14.36 lbs/gal
Slurry yield = 1.84 cu ft/sack
Seawater = 9.1 gal/sack
Lead slurry volume: 2656 − 125 = 2531 cu ft.
2531 cu ft requires:
1376 sacks class G
12,522 gal salt water
34,696 lb salt
2103 lb Attapulgite
Total requirements for 9⅝″ casing:
Class G = 2334 sacks
Attapulgite = 2952 lb
Salt = 56,408 lb
Seawater = 20,371 gal

4. 7″ liner 13,608± ft to 11,126 ft = 2482′ with 200 ft in 9⅝″

2482 − 200 = 2282′ open hole; hole size 8½″
Recommend class E cement mixed with 18% seawater
Improve flow properties with 0.75% CFR to induce turbulance.

Slurry yield	= 1.09 cu ft/sack	
Slurry weight	= 16.7 lb/gal	
Open hole volume		
2282 × 0.1268	=	289 cu ft
Annular volume 200 × 0.1438	=	29 cu ft
		318 cu ft

Requirements for 318 cu ft are:
- 294 sacks class E cement
- 1263 gal seawater
- 1576 lb salt
- 193 lb CFR-2

7.11 Alternative Casing Programs

A.

Casing	*Depth*	*Type*
30″	Preparation driving 400 ft	30″ × 1″ W.T. grade B
20″	1620′	94 # H-40 ST&C
13⅜″	Top 200′	54.5 # S-80 ST&C
	1200′	61 # S-80 ST&C
	1600′	68 # S-80 ST&C
	2700′	72 # S-80 ST&C
	2076′	81.4 # S-95 ST&C
	7776′	
9⅝″	Top 5500′	40 # N-80 LT&C
	2200′	40 # S-95 LT&C
	2700′	43.5 # S-95 LT&C
	926′	47 # S-95 LT&C
	11,326′	

	7" (11,126 to 13,608')	2482'	26 # S-95 LT&C
	7" top to bottom	2200'	23 # N-80 Buttress
		3800'	23 # N-80 LT&C
		3700'	23 # S-95 LT&C
		3908'	26 # S-95 LT&C
		13,608'	
B.	20"	1620'	133 # K-55 Buttress
	13⅜"	Top 4500'	72 # N-80 ST&C
		1500'	72 # S-95 ST&C
		1776'	81.4 # S-95 ST&C
		7776'	
	9⅝"	Top 1700'	40 # N-80
		3100'	40 # N-80 LT&C
		2200'	40 # S-95 LT&C
		2800'	43.5 # S-95 LT&C
		1526'	47 # S-95 LT&C
		11,326'	
	7" top to bottom	3500'	23 # N-80 Buttress
		2200'	23 # N-80 LT&C
		3700'	23 # S-95 LT&C
		4208'	26 # S-95 LT&C
		13,608'	

7.12 Drilling Curve

The drilling curve depicted in figure 7.3 indicates the expected drilling time for the well with the recommended parameters, equipment, and proper supervision. It includes expected times for running various casing strings as well as 10-day testing. Total time (including the 10-day testing) for regular drilling is

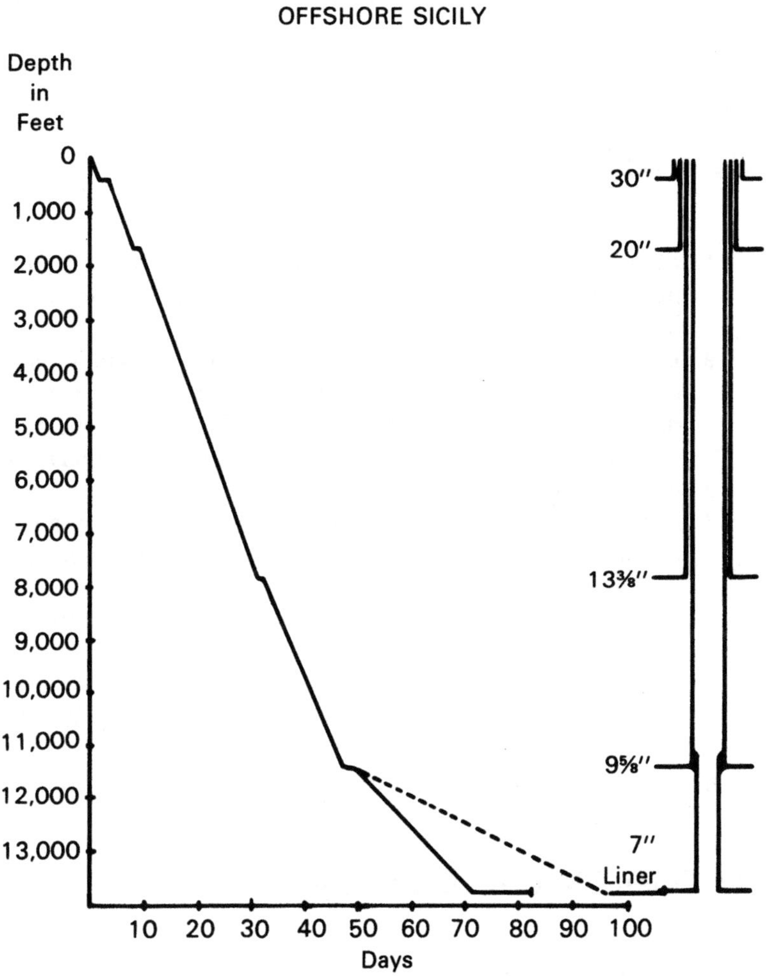

Figure 7.3 Drilling curve.

expected to be 82 days. Should you elect to core the bottom interval, the total time would be extended to 117 days.

The curve is an estimate with no allowances made for lost circulation, waiting, on orders, supplies, weather or mechanical difficulties.

7.13 Logging Program

1. Should logs be run prior to productive interval, it would be helpful to have caliper, gamma-ray, cement-bond logs.
2. Logs recommended in the productive interval are: gamma-ray, compensated-neutron formation density, dual laterolog, microlaterolog, borehole compensated sonic log. The borehole compensated sonic should have 3-in. spacing to help define fracture area.
3. Borehole should be clean with drilling fluids in good condition. There should be proper mud weight and the viscosity should be near 60.
4. Logging tools should be checked prior to lowering in the well to insure minimum open-hole condition.
5. Should logging operation require a long interval of time, the drilling supervisor should consider a short trip in hole to condition mud.
6. Keep wellbore full of drilling fluids during logging operations.

7.14 Open-Hole Well Testing

1. Caution should be exercised in all open-hole testing. See figure 7.4 for tool assembly.
2. Tests should be run in daylight hours only. Routine safety drills should be run prior to test.
3. Tests should be scheduled so as to insure enough daylight

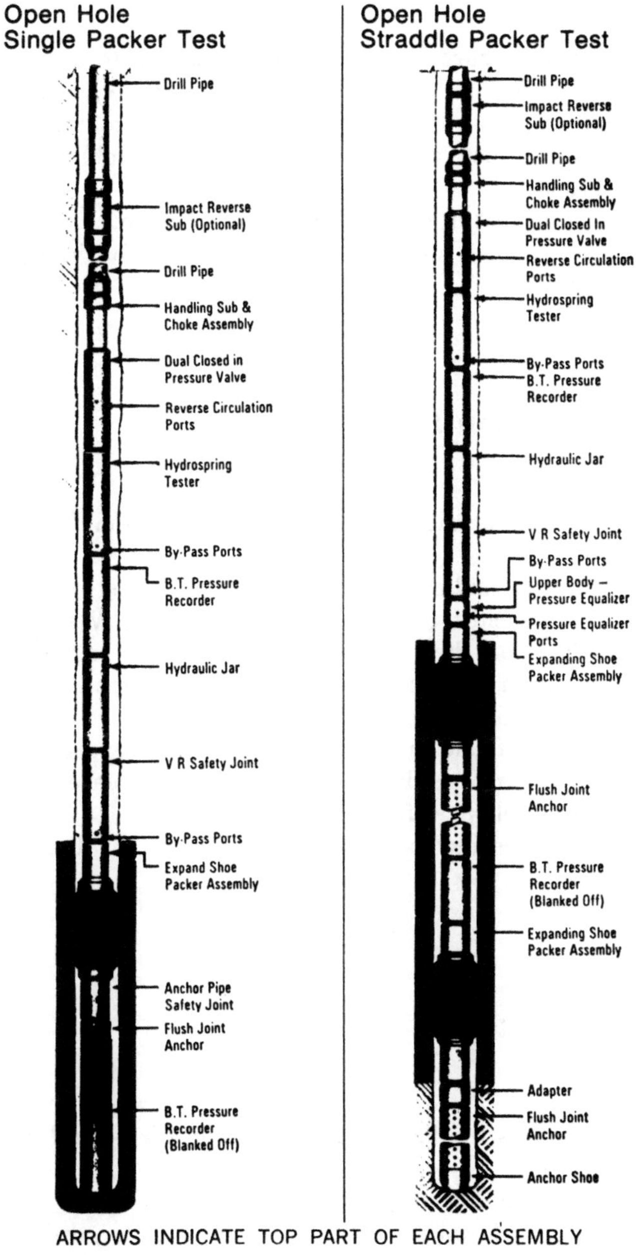

Figure 7.4 Typical PST assemblies. (Courtesy of Halliburton Services)

hours to secure adequate data, that is, 3 to 5 min flow, shut-in for 30 min, flow test for 4 hours, shut-in for buildup 4 to 5 hours (total time 10 to 11 hours).

4. Conduct test after pulling core to establish gas/oil contact.
5. Conduct test in middle interval of oil zone.
6. Conduct test after pulling core to establish oil/water contact.

7.15 Completion Program

1. Perforate, acidize, and test wellbore from bottom to top.
2. Perforate, acidize, and test small intervals (10± meters).
3. Use frac-acid treatment instead of matrix. (Wells drilled in tight or dirty formations may show little or no production on drillstem tests but may be good producers after fracturing.)
4. Injection of acid in 2⅞" tubing should not exceed 4 BPM or 1656 psig. Additional rates would increase pressure and possibly acid-frac downward.
5. Monitor casing pressure continuously for leakage of packer.

7.16 Completion

1. Test rams, choke, and kill lines to 5000 psi and BOP to 2500 psi.
2. Condition fluid in pits to 10 #/gal.
3. Run in hole with 6⅛" bit, 7" casing scraper, drill collars and drillpipe to top of liner—do not use jets in bit.

4. At top of liner, condition fluid in hole so that the system is balanced.
5. Proceed into 7″ liner carefully and clean out to total depth, circulate wellbore clean. Spot acid 15% NE HCI to cover perforating interval.
6. RU perforators—RIH with GR/collar locator log. Correlate perforations with previously run logs. Perforate the proper intervals by using a 4″ OD steel hollow carrier, decentralized select-fire gun, 19 gram charge, 0.46 entry hole, two perforations per foot of perforating interval. Perforate top to bottom. Pull out of hole.
7. Pick up 2⅞″ 6.50 lbs N-80 tubing string with straddle packer assembly for 7″ casing. Drift each joint during pick up with 2.165″ diameter drift. Pressure-test each threaded connection to 3500 psi while running string with Gator Hawk unit. The minimum collapse for 2⅞″ N-80 tubing is 9420 psi.
8. Run SPA to lowest perforations; spot acid to packer and treat the perforated intervals.
9. Pull up hole with straddle packer assembly to give working room above top acidized perforation. Set packer and RU swab. Swab load acid back collecting water samples to check pH and chlorides.
10. Swab-test acidized perforations to determine productivity of zone.
11. Do not swab at night and during flow-test or swab-test. Burners and test unit must be manned.
12. Flow test well until stabilized, then have a 4-hour flow period to determine potential on minimum, medium, and maximum chokes.
13. After three flow-rate tests, shut well in for bottomhole pressure survey.
14. When tests have been completed, oil phase fluid down tubing, open bypass, and pull straddle packer assembly.
15. Run in hole with production packer and set with good measurements.
16. If subsea completion, pull bore protector getting good

measurement on overpull when retrieving tools lands and mark guide lines for tide diffference.

17. Install special slotted protector, nipple above and below the DHSV nipple. As you go in with ball-valve receptacle, do not attach steel control line on space out run.
18. Space out by stabbing into packer with respect to a previously fixed point on the guide line. Pick string back up to ball-valve receptacle, attach ¼" steel line and go back in hole.
19. Connect ¼" control line to DHSV-1 landing nipple, test to 6000 psi.
20. Install tubing hanger on tubing.
21. Connect ¼" tubing DSV-1 port on tubing hanger, install tubing hanger, reentry tool, and flush DSV-port.
22. Pull separation sleeve above DSV-1 landing nipple, flush ¼" line with hydraulic oil. Reinstall separation sleeve in DSV-1 landing nipple. Test DSV-1 control line with 6000 psi. Remove reentry tool.
23. Land tubing in wellhead, land reentry tool, and rig up wireline equipment.
24. Set check valve in packer tail pipe, do not release from plug.
25. Test tubing with 3500 psi, monitor annulus pressure, release pressure.
26. Test tubing with 2000 psi, monitor tubing pressure, release retrieve check valve.
27. Set blanking plug in landing nipple in tubing below seal assembly test tubing and plug with 2000 psi.
28. Set separation sleeve in DHSV landing nipple and test control line to 4000 psi.
29. Set blanking plug in landing nipple, test above plug with 3500 psi, monitor annulus.
30. Displace riser through choke and kill lines using seawater, pull riser.
31. Pull BOP, move rig off well as soon as stack is clear of guide base.
32. Move rig back over wellhead, have divers inspect wellhead tubing hanger. Move tree under rotary table and prepare for running.

33. Install tree and test.
34. Pull blanking plug. Displace tubing with diesel.
35. Connect flow line and pressure test production equipment.
36. Open well and flow to clean up, test as required, monitor annulus pressure while flowing; pressure will increase due to temperature expansion. Allow pressure to increase to 1000 psi, then bleed back to 500 psi.
37. Displace riser and tubing to DHSV landing nipple with diesel, close DHSV valve, bleed pressure off tubing, observe 30 min. Repressure tubing and open downhole safety valve.
38. Close swab valve, observe well for 30 min, close master valve, annulus, and valve, and DHSV.
39. Release all hydraulic pressure from tree. Pull and lay down riser.

A testing procedure for well test along with bottomhole pressure buildup is shown in figure 7.5, with the form for production test calculations shown in figure 7.6.

7.17 Abandonment

In the event of abandonment, the following general procedures will be used.

1. Open-hole abandonment:
    a) Run 100 ft plug on bottom; no feel is required.
    b) Run 500 ft plug over or between all porous intervals of different geological ages; feel plugs after 8 hours.
2. Cased-hole abandonment:
    a) Run wire line, set bridge plug at base of 9⅝″ casing. Pressure test to 1500 psi, then place 25 ft cement plug on top.
    b) Run bridge plug at 200 ft below seafloor and dump 25 ft of cement on top of plug.
    c) Remove all tubulars and seafloor equipment.

Well No:________________Zone:______________ZERO correction______

Procedure

1. Rig up lubricator and tools.
2. Shut in well, install swab valve and lubricator.
3. Retrieve subsurface safety valve at____________means depth.
4. Run in hole with____________________ psi bomb and________ hour clock and thermometer.
5. Hang bomb in___________________ at ____________means depth.
6. Pull out of hole, install subsurface safety valve, and return well to production.
7. Allow well to stabilize for a time equal to approximately 8 times the duration of the well being shut in to run bombs. (Maximum flowing time not to exceed 24 hours.)
8. During the last six hours of flowing period, obtain a well test consisting of hourly oil, gas, and water rates. Read tubing pressure, manifold pressure, separator pressure, and temperature and choke size at last hour of test. Collect water-cut and oil-gravity samples during first and last hour of test.
9. At completion of test, shut in well at wing valve and obtain a buildup of _________hours.
10. After buildup, pull subsurface safety valve, retrieve bomb.
11. Obtain gradient stops at test depth, 200 ft above test depth, and every 1000 ft thereafter.
12. Install subsurface safety valve, rig down lubricator, and swab valve.
13. Advise production supervisor that well is ready to be returned to production.

Figure 7.5 Testing procedure.

7.18 Blowout Prevention Procedures

These are orders issued by all supervisors on the rig; contractor and personnel representing the company are directed to follow these procedures. No hole is to be drilled and all operations must cease if procedures are not in full effect and serving the obviously intended purpose.

These procedures are to be thoroughly discussed between each Drilling Superintendent, Drilling Supervisor, Tool Pusher, and Driller in a special meeting held solely for this purpose. A copy of the list of practices is also to be handed by the chief supervisor present to each of these supervisors.

Failure to carry out this directive in any part by any supervisor shall be treated as the most serious breach of job conduct. Checks are to be made during routine and special operations. Punitive action (if necessary) shall be taken on the spot—before a blowout occurs.

1. Pit drills: One or more shall be conducted each tour.
2. Inside BOP drills: One or more shall be conducted each trip.
3. Accumulator drills: One or more shall be conducted each week.
4. Hole fillup: Hole is to be filled on trips while pulling out of hole at two-stand intervals. (Keep full also while GIH with drillpipe and casing). Run it over. Check to be certain that the hole is taking the correct amount of mud. Normally, drillpipe is to be slugged prior to pulling out of hole.
5. BOP equipment installation: All BOP and lines, the accumulator unit, the remote controls, the pit-level indicator, all wellhead and BOP valves, and all other parts of the well-control system are to be installed, operative, and in good condition when open hole exists. Pit-level indicator is to be positioned directly in front of the driller's position.
6. Position of BOP-actuating valves: All three (or four) BOP actuating valves on the accumulator unit are to be

PRODUCTION TEST CALCULATION PROCEDURE

WELL NO.	DATE CHOKE HOURS	METERED (1) GROSS BOPD* (METER)	(2) BWPD (METER)	% BS & W (3) IN GROSS BOPD	BWPD (4) IN GROSS BOPD (1)X(3)	TOTAL (5) BWPD (2)+(4)	(6) BOPD (1)-(4)	(7) BFPD (5)+(6)	% (8) WATER (5)÷(7)	(9) MCF/D	(10) COR (9)÷(6)	(11) FLOWING PRESSURE	(12) API	(13)

* CORRECTED FOR METER FACTOR DETERMINED BY ACTUAL GAUGED AMOUNT VERSUS METER READING

Figure 7.6 Production test calculation procedure.

kept in BOP "open" position. This eliminates the possibility of turning the valve handle in the wrong direction. This is a must, otherwise there will be no pressure "on" the seals of hydrill, and mud can enter the oil system. Valves can break and BOP will creep shut if handles are in neutral position on console.

7. Position of BOP side-outlet valves: All side-outlet and other valves in the entire system, including the choke manifold, are to be kept closed. The only exception to this is during special well-control situations—and then only during the current emergency period, in accordance with the supervisor's judgment.
8. Driller's check on (6) and (7) above: Driller will check all valves prior to relieving man on duty. Tool Pusher shall occasionally open (or incorrectly position) a valve and observe whether the Driller discovers this. Following this check, the valve will be repositioned.
9. BOP and accumulator operation: Actuate BOP rams from main accumulator station every trip. This proves the performance of the preventer and the accumulator.
10. Remote controls: Actuate BOP rams from remote station every trip. This proves the performance of the remote control.
11. Inside BOP position: Stand upright in front of draw-works, keep set in "open" position. Have three handles, which are 1½′ long, installed on sides of upper removable piece. Be sure that there are two handles (one for the drillpipe and one for drill collars) or an equivalent plan that uses one valve with appropriate subs.
12. BOP tests: Test the entire system including all preventors and lines to the rated working pressure (except Hydrill 2500 psi) on each nipple-up and on trips after each 48-hour drilling period. (Do not pull out just to test, but rather perform the test during the trip for bit change, which arises after the 48 hours of on-bottomhole making.)
13. Casing tests: Test entire casing string (70% of rated burst press or at working press of BOP, whichever is

less). After each casing setting, test casing on trips after 48 hours of drilling (1500 psi—13⅜"; 2500 psi—9⅝"; 3000 psi—7"). the 48-hour test is to be against a cup positioned in casing 30′ to 90′ below casinghead. This test must be made again just prior to setting subsequent strings of casing. If a leak is found, it is to be corrected by replacement, if possible, or by filling annulus with cement, or both.

14. Fill casing: Casing is to be kept approximately full while running. Completely fill at least every five joints.
15. Weighing mud: The mud going into the drillpipe and mud coming out of the hole is to be weighed each 30 minutes while drilling. These measurements are to be clearly recorded in table form on a prepared sheet of paper. File these records with company drill reports (morning reports). Changes in mud weight of as much as 2 lb/gal are to be immediately reported to the driller. The mud-weighing person will keep an eye on the flowline and pits and will advise the Driller of abnormal discharge of liquids from the hole, while drilling, circulating, during connections, and at other times if present.
16. Use of drillpipe float: Under normal circumstances, run a back-pressure valve (Baker type, for example) to 6,000. This is not an inflexible requirement. If the Tool Pusher finds reason to discontinue the use of the float, he may do so.
17. Accumulator cleanness: Keep the entire unit (including valves, frames, piping, pumps, and regulators) clean and in particular keep it free of rags, wrenches, soft line, rope, gloves, loose-fitting raincoats, hats and other objects.
18. Accumulator and remote-control valves and labels:
    a) Keep the BOP closing valves arranged in same order as the BOP, i.e., first, the Hydrill, second, the top-pipe rams, third, the blind rams, fourth, and bottom-pipe rams. (Where only two sets of rams are installed, the pipe rams normally shall be on bottom and the blind rams on top.)
    b) Keep the labels *Hydrill*, *Top-Pipe Rams*, *Blind*

Rams, and *Bottom-Pipe Rams* legible by keeping them clear of smears of paint, grease, dirt, or other material.

19. Hydraulic fluid: DO NOT use pure diesel oil or other fluids in the accumulator, which can damage rubber parts of the BOP equipment. Light lube oil (Gulf 372, for example) or preferably noncombustible hydraulic fluid is recommended.
20. Plugged lines: Pump through all lines once each day (report Friday) or more often if necessary to be sure that they are not plugged.
21. Plugged degasser: When a degasser is installed on the rig, it is to be checked at the beginning of each well or more often to make sure that it is not stuffed up with shale, mud, solids, etc. Do this before spudding and record on the International Association of Drilling Contractors drill report.
22. Fillup line: The fillup line is to be connected to the bell nipple.
23. Use of large Hydrill:
    a) The following rules apply to certain special-type wells, particularly (but not limited) to offshore and deep wildcats. A type G.K. Hydrill BOP is to be correctly installed on all drive pipes, and conductor pipes and tested. As a further general guide, the Hydrill is to be used at all drilling depths over 500′ (or lesser depths under many circumstances) when casing has been set. If 20″ or approximately this size conductor is not used, the Hydrill will be installed on the drive pipe.
    b) Do not remove the Hydrill from conductor pipe after setting surface casing until cement in the upper annular space between the conductor pipe and surface casing is full of SET cement. Furthermore, do not cut holes in the conductor or otherwise open and expose the well until the cement is set. Do the recementing (if cement did not circulate) with pipe lowered through the Hydrill.

c) Two 4-in. (or 6-in.) steel lines will be extended from immediately below the 20′ Hydrill to a safe distance away from the rig structure. These are to relieve gas flows which cannot be contained by shallow casing strings. The lines are to be connected to the well by 4-in. or larger valves.

24. Cement jobs during nipple-up: Careful attention must be given to cementing the upper part of shallow casing strings. The upper annular spacer between the surface casing and conductor pipe and space between conductor pipe and drive pipe (or other similar casing) must be completely cemented, either by circulating cement to the surface (preferred) or by recementing this space through small pipe inserted from the top. This action applies under all circumstances (with the possible exception of floating rigs) and to any type wellhead equipment used (except subsea equipment in very deep water). After getting a cement job at the top, check it again to see that the cement is hard and that it has not dropped. If hard cement is not present all the way to the top, recement the casing. DO NOT DRILL UNTIL IT IS FULL OF SET CEMENT. Weld steel plates between the bottom edge of the braden head and upper-conductor casing. Install valves in the plates so that the cement level may be checked with a rod. The importance of this procedure must not be underestimated. Cement is needed for structural support to help prevent vibration cracks and leaks in the casing threads and all other upper parts of the well.

25. Wellheads:

a) At the time of each and every nipple-up operation, all wellhead seals and connections will be pressure-tested to their rated working pressure. Blowout-preventer tests also will include testing the wellhead seals (rated working pressure).

b) All outlets on wellheads will be equipped with flange-fitted valves of maximum safe working pressure. It is preferred that two valves be attached in series with flanges to each outlet.

Immediately prior to moving the rig off a completed well, the outermost valves will be plugged with bull plug. The bull plug will be tapped and threaded with pipe threads, and a gauge cock will be screwed into the threads. Provision should be made for use of a valve-removal plug in each flanged outlet (this is optional, but encouraged). If only one valve happens to be present on the outlet, this is required. One pressure gauge is to be installed in one gauge cock on each string of casing.

c) If drilling is done near a cluster of wells, such as on an offshore platform, the drilling crew must check and record on the drill report the pressures on each string of casing each day. Should the pressures become abnormally high, this must be called to the attention of higher authority.

26. Well completions: Upon completion of every well, whether it is a new well or workover of an old one or one that has been plugged and abandoned, the contract Tool Pusher shall personally inspect the wellhead to determine that it is safe to leave the well permanently unattended. He will check all flanges by using a wrench to see that all nuts and bolts are tight. The Tool Pusher will check all valves (to see that they are closed and sealed in accordance with other instructions herein) and make other examinations including determining if casing annulus is full of cement. This will be done with the Tool Pusher's own hands. The task will NOT be assigned to a subordinate. He will then write on the IADC drill report that the inspection has been made. This record will be in longhand in the Tool Pusher's own handwriting (see fig. 7.7). The Drilling Supervisor will check to see that the Tool Pusher carried out this assignment. If at all possible, the company representative should witness the inspection on at least selected locations. Should unsafe conditions be found, they must be corrected prior to taking the rig apart, no matter how long it takes.

(Write this on IADC Report - in longhand writing of the person who made the inspection.)

STATEMENT OF FINAL INSPECTION OF ALL WELLS

I,____ have personally inspected both visually and with my own hands all NUTS, VALVES, and FLANGES on well ____
number lease

Area at _________________________Hours, _________________________
date

and know that all valves are closed, all flanges are tight, and this well is safe in every respect to move off, to the best of my knowledge. I also know that all tubular members (drivepipe, conductor pipe, etc.) outside the surface casing have been rechecked to make certain they are filled to the top with cement. All valves and fittings installed on the well are in accordance with the desired method for leaving them.
Casing pressures are as follows__________________________________
________________________.

SIGNED________________________ SIGNED________________________
Company Supervisor Contract Tool Pusher

DATE________________________ DATE________________________

Figure 7.7 Statement of final inspection of wells.

27. Wireline Work:
    a) Lubricators used in "piano wire" work (for example 0.082″ or 0.090″) and during jobs requiring wires of other sizes will be pressure-tested to the rated working pressure of the lubricators each time they are used on a well when the rig is on location. Check each lubricator to be certain that it is strong enough. Determine the working pressure of the lubricator—at the weakest point. Always have something that will screw into anything you are going to run in hole to shut in the well. To have a tubing valve always on floor beside the Driller is a must.
    b) Wireline lubricators are encouraged when wirelines are run under most well conditions. It is usually obvious to the Driller when a lubricator should be used, and no chances should be taken with it.
28. Barite stocks: A minimum of 3,000 cu ft of barite (but preferably 5,000 cu ft, space permitting) will be kept at some place either on the rig, near the rig, or within a reasonable distance from the rig. Also, keep enough gel and chemicals (including salt, if needed) to match. Additional stocks of much larger volume should be available within the country of most operations (500 to 1,000 tons).
29. Abandon-ship drills: On any offshore rig, one or more drills are to be conducted on the first day of each week and reported every Friday morning. This applies to all floaters and all bottom-supported rigs.
30. Smoking: Smoking shall be permitted only in designated areas, such as a doghouse outside rig-floor and mud pit area, in the living quarters, and inside mud logger unit and slumberger unit. No matches, lighters, cigarettes, cigars, or pipes are to be carried in close proximity to the rig.
31. Plug and abandonment procedures:
    a) All open-hole shows (presumably thin noncommercial stringers or other unwanted pay zones)

are to be sealed by spotting an open-hole cement plug, not less than 300′ in length (100′ to be below the show and 200′+ above the show). In no case is the plug to be less than 100 cu ft in volume of cement. Check plug with bit.

b) A cast-iron cement retainer is to be set as deep as practical (preferably about 50′ from bottom) in each unplugged string of casing left in the hole. No less than 50 sacks of cement will be bumped out below the retainer and approximately 100+ sacks dropped on top. The plugged casing will then be pressured-tested to 1500+ psi. On land wells, a steel cap and valve are to be welded on top of the casing stub, after the casing has been cut off below the "plow point" subsurface depth. Other special action is to be taken on offshore jobs, preferably planned well in advance of the abandonment. This includes blasting or cutting off the well below the ocean floor and removing of major debris.

32. Reporting: The fact that these procedures are (or are not) being followed exactly as directed is to be reported at least once a week (at the end of each week) to the local Company Manager and the Drilling Contractor's local Manager.

7.19 Pit Drill

The purpose of the pit drill is to train the Driller to be constantly aware of the fluid level in mud pits (much like the driver of an automobile subconsciously checks his speedometer, out of habit, to know how fast he or she is going). This training is expected to prepare the Driller to detect threatening blowouts at the first surface indication and take preventive action soon enough to avoid disaster.

The pit drills are to be conducted under direction of the Tool

Pusher. One or more pit drills shall be held each tour. This is the minimum requirement under company policy. Several (6 to 8) drills each tour should be given to Drillers who have not been thoroughly trained in this practice.

No advance notice will be given to the Driller that a drill is to be conducted. The idea is to test the degree of vigilance exercised by the Driller over this vital indication of impending mass destruction.

Drills are to be conducted during both routine and special operations. For example, they will be given while drilling, when the rig is down due to equipment repairs, while logging, waiting on orders, circulating, when the Driller has gone to eat and is replaced by one of his assistants, when the Driller is talking to someone, or at any other time when there is an open hole and blowout preventers are installed.

All equipment required for the drill is to be installed prior to drilling. It is to be kept in good operating condition. If necessary, any required equipment will be obtained by air freight.

A Pit-O-Graph or its equivalent shall be obtained (by urgent air freight, if not available on rig) for use at all times on each well. In the event that this system is not available or is malfunctioning or otherwise becomes inoperative, a homemade pit level indicator will be rigged up with float, cable, pulleys, and an indicator installed in front of the Driller's position at the drawworks.

Pit-O-Graph charts with recorded pit drills, hole fill-up practice, etc. shall be filed with Company drill reports. Drillers will note time of each drill (in minutes) on each chart. The charts will be signed by the Tool Pusher and the Driller.

The fact that pit drills were held will be announced each morning on both the verbal-drill and the IADC report. Number of drills conducted and reaction time of each will be registered. This is to be in writing on the morning-report form which is kept in the Company office. The procedure shall be as follows:

1. The float in the mud pit shall be raised (or lowered). This indicates that the well is coming in (or losing returns).
2. The Driller will notice the change in pit level.
3. The Driller will indicate this by giving an audio and visual signal. Normally, this is done by blowing a horn or siren

immediately and pulling up the kelly if drilling. The Driller will then check, rapidly, to see whether the well is flowing, and act accordingly. NOTE: In the absence of a horn, the Driller may shout: "Blowout!"

4. The Driller's reaction time from the point when the float is first raised until the signal is given will be checked with a stop-watch.
5. The reaction time for each drill shall be recorded on the Pit-O-Graph chart and the drill report.
6. Acceptable reaction time is 1/2 minute. But, preferably, the Driller should catch the change quicker—almost instantly. Important NOTE TO DRILLERS: If the well starts flowing, do this:

    STOP THE PUMP.

    SHUT THE WELL IN, QUICKLY.

    THEN DO WHAT IS NECESSARY, INCLUDING CALLING THE PUSHER.

7.20 Inside BOP Drill

The inside BOP drill serves two important purposes. First, it gives practice and trains the Driller and the crew to close-in the well quickly. It is desired that they become experts in this performance. Second, it proves that essential equipment is installed and is in good operating condition.

One or more drills will be given during each trip (while pulling out of the hole and going into the hole) after the bit is up in the casing. One drill each trip is minimum company policy.

The inside BOP drill will be supervised by the Tool Pusher. All parts of the well-control system will be kept hooked up in good condition and ready for drills at all times.

Regular drills will commence after the installation of the first blowout preventer (20" Hydrill or 12" BOP, etc.).

In effect, a simulated blowout will be experienced by the crew on each trip. They will learn to shut-in the well "automatically" without having to think about how to do so. The procedure shall be as follows:

1. The float in the mud pit shall be raised. This shows (on the indicator in front of the Driller) that the well is starting to blowout.
2. The Driller will notice the change in pit level. He won't stop to see if this is a drill—he assumes this to be the real thing.
3. He immediately will sound the alarm to his men—by blowing his horn or siren and shouting "Blowout." All hands will act accordingly.
4. The crew will shut-in the well as quickly as possible:
   a) The top set of pipe rams (Hydrill, in case collars are up) will be closed.
   b) The inside BOP will be installed in the drillpipe or collars and closed. Note that equipment (threads) must be available to shut in the well when any size drillcollar is on the rotary table.

   Note: Reverse order of (b) and (a) if drillpipe float is not in service.
5. The total time elapsed from the point when the float is first raised until the well is shut-in will be checked with a stop-watch.
6. The passing time for the test is 1½ minutes. Top-notch crews may learn to do it quicker.

7.21 Accumulator Drill

Accumulator performance will be proven at the time of installation of blowout preventers. Thereafter, this drill will be conducted one or more times each week. Results of drills, number held, closing times of the rams and Hydrill, and initial and final accumulator pressures are to be reported on the Friday-morning drill report.

All accumulator drills will be conducted when drillpipe is not in open hole (up in casing—normally out of the hole). The Tool Pusher will witness and supervise the drills. The procedure shall be as follows:

1. The accumulator pumps will be turned off.
2. Initial accumulator pressure will be recorded.
3. All preventers will be closed at same time.
4. Closing time for each preventer will be checked with a stop-watch.
5. Preventer-closing times will be recorded.
6. Final accumulator pressure will be recorded.
7. To pass the test, ALL BOP must be closed in less than 20 seconds and final accumulator pressure must be 1150 psi or greater.
8. Accumulator pumps will be turned back on. NOTE: Drill is to be conducted every other week from the remote controls. If the equipment does not meet the requirements indicated in point (7) above, IT IS NOT IN GOOD CONDITION. IT MUST BE IN GOOD CONDITION. FIX IT IF IT NEEDS FIXING—RIGHT THEN!

Appendix A

The following tables would be used to fill out the Dresser Industries hydraulic worksheet. The information in this appendix is reprinted courtesy of the Security Division of Dresser Industries.

Table A.1: Contains the pressure ratings (in psi and the volumetric discharge (gal/stroke).

Table A.2: With the pump output (gal/stroke) from table A.1 and the pump rate (stroke/min) utilized, obtain the circulation rate (gal/min). Circulation rates are based on 90% volumetric efficiency for duplex pumps.

Table A.3a: The annular velocity depends on the circulation rate, hole or casing size, and size of the drillpipe. Using the circulation rate from table A.2 and the hole and pipe size, calculate the annular velocity around the drillpipe. This annular velocity should equal or slight exceed the value established as the minimum annular velocity around the drillpipe.

Table A.3b: Using table A.2, circulation rate with the hole and drillcollar size, determine the annular velocity around the drillcollars.

Table A.4: Describes four types of surface equipment (standpipe, hose, swivel, and kelly). Select the type that describes the surface equipment best.

Table A.5: Determine the pressure loss through the surface equipment by correlating the surface equipment type (table A.4) with the circulation rate (table A.2).

Table A.6: Using the circulation rate and the drillpipe size, weight and type, determine the pressure loss through the drillpipe bore. Note that the loss is in psi/1000 ft. The pressure loss for the entire drillpipe string is calculated as follows:

$$\text{Total loss} = \frac{(\text{Press. loss from Table A.6})}{1{,}000} \times (\text{Length of drillpipe}).$$

Table A.7: Using circulation rate, the hole size and drillpipe size, determine the pressure loss in the drillpipe annulus. Note that the pressure loss is given in psi/100 ft. Total loss can be calculated as in table A.6.

Table A.8: The circulation rate and the drillcollar bore size determine the pressure loss through the drillcollar bore. Note that the loss given by the formula above is in psi per 100 ft.

Table A.9: Using the circulation rate, hole size, and drill-collar size, determine the pressure loss in the drillcollar annulus. Note that the loss given in table A.9 is in psi 100 ft.

Table A.10: The pressure available for nozzle selection is the difference between the operating-pressure limit and the actual system-pressure loss, corrected to 10 ppg mud. Table A.10 lists pressure losses through the nozzles and the jet-nozzle area for different combinations of parameters. Using the circulation rate, select a jet-nozzle size combination for which the pressure loss is equal to or less than the amount of pressure available. For step 11 of the worksheet this jet nozzle selection should yield optimum energy expenditure at the bit for a given circulation rate and pressure limit. The greatest jet-impact force is delivered if the pressure drop across the bit nozzles is 48% of the surface pressure.

Table A.11: Obtain the jet velocity by using the nozzle combination from table A.10 and the circulation rate.

TABLE A.1 PUMP DISCHARGE PRESSURE (psi); PUMP DISCHARGE VOLUME (gal./stroke)

Manufacturer: GARDNER-DENVER – Triplex

Model	Max I.H.P.	Max S.P.M.	Stroke Length	Liner Size (in) 3	3-1/4	3-1/2	4	4-1/2	5	5-1/2	6	6-1/4	6-1/2	7
PJ-8	275	175	8	3118 0.8	2657 0.8	2290 1.0	1753 1.3	1386 1.6	1122 2.0	– –	– –	– –	– –	– –
PY-7	500	160	7	– –	– –	– –	– –	3150 1.4	2550 1.8	2110 2.2	1770 2.6	– –	1510 3.0	1300 3.5
PZ-8	750	165	8	– –	– –	– –	5381 1.3	4238 1.6	3433 2.0	2843 2.5	2385 2.9	2200 3.2	– –	– –
PZ-9	1000	150	9	– –	– –	– –	– –	5530 1.9	4485 2.3	3710 2.8	3110 3.3	2875 3.6	2650 3.9	2285 4.5
PZ-11	1600	130	11	– –	– –	– –	– –	– –	– –	5595 3.4	4702 4.0	– –	4006 4.7	3454 5.5

TABLE A.1 *(continued)*

Manufacturer: **IDECO – Duplex**

Model	Max I.H.P.	Max S.P.M.	Stroke Length	Rod Size	Liner Size (in) 3-3/4	4	4-1/2	4-3/4	5	5-1/4	5-1/2	5-3/4	6	6-1/4	6-1/2	6-3/4	7	7-1/4	7-1/2	7-3/4	8
MM-200	200	80	10	1-7/8	2000	1825	1460	–	1163	–	970	864	790	–	667	617	–	–	–	–	–
					1.7	2.0	2.5	–	3.1	–	3.8	4.2	4.6	–	5.5	5.9	–	–	–	–	–
MM-300	300	80	12	2	2500	2380	1830	–	1458	–	1185	–	985	–	832	–	712	662	–	–	–
MM-300GB					2.0	2.3	3.0	–	3.7	–	4.6	–	5.5	–	6.6	–	7.7	8.2	–	–	–
MM-450	450	80	12	2-1/4	–	–	2830	2510	2225	–	1810	–	1500	–	1265	1165	1082	1000	–	–	–
					–	–	2.8	3.2	3.7	–	4.5	–	5.5	–	6.5	7.0	7.6	8.2	–	–	–
MM-550	550	65	15	2-1/2	–	–	–	3120	2775	2480	2235	2020	1845	1690	1550	1425	1320	1220	–	–	–
MM-550F					–	–	–	4.0	4.5	5.0	5.5	6.1	6.7	7.3	8.0	8.7	9.4	10.1	–	–	–
MM-600	600	65	16	2-1/2	–	–	–	–	2830	2540	2280	2060	1880	–	1582	1348	1250	1165	–	–	–
					–	–	–	–	4.8	5.3	5.9	6.5	7.1	–	8.5	9.2	10.0	10.8	–	–	–
MM-700	700	65	16	2-3/4	–	–	–	–	–	3038	2730	2470	2246	–	1878	–	1595	1487	1375	1285	–
MM-700F					–	–	–	–	–	5.2	5.8	6.3	7.0	–	8.4	–	9.9	10.6	11.4	12.2	–
MM-900	900	65	16	3	–	–	–	–	–	–	3810	3250	2950	–	2459	–	2085	1933	1795	1670	1562
					–	–	–	–	–	–	5.2	6.0	6.8	–	8.2	–	9.7	10.4	11.2	12.1	12.9
MM-1000	1000	65	16	3	–	–	–	–	–	–	4020	–	3280	–	2735	2510	2325	2155	2000	1860	1740
MM-1000GB					–	–	–	–	–	–	5.2	–	6.8	–	8.2	8.9	9.7	10.4	11.2	12.1	12.9
MM-1250	1250	65	18	3-1/8	–	–	–	–	–	–	–	–	3680	3350	3065	2820	2600	2400	2230	2079	1940
					–	–	–	–	–	–	–	–	7.6	8.4	9.2	10.0	10.8	11.7	12.6	13.5	14.5
MM-1450F	1450	65	18	3-1/8	–	–	–	–	–	–	–	–	4270	3880	3560	3270	3010	2790	2590	–	–
					–	–	–	–	–	–	–	–	7.6	8.4	9.1	10.0	10.8	11.7	12.5	–	–
MM-1625	1625	65	18	3-3/8	–	–	–	–	–	–	–	–	4920	–	4060	3790	3430	3170	2940	–	–
					–	–	–	–	–	–	–	–	7.4	–	8.9	9.7	10.6	11.5	12.4	–	–
MM-1750F	1750	65	18	3-3/8	–	–	–	–	–	–	–	–	5000	4800	4380	4020	3700	3410	3175	–	–
					–	–	–	–	–	–	–	–	7.4	8.2	9.0	9.8	10.6	11.5	12.3	–	–

NOTE: Top Value: Discharge Pressure / Bottom Value: Discharge Volume

DRESSER

TABLE A.2 CIRCULATION RATE; DUPLEX PUMPS (gpm)

Strokes/Minute	GALLONS PER STROKE																	
	6.2	6.3	6.4	6.5	6.6	6.7	6.8	6.9	7.0	7.1	7.2	7.3	7.4	7.5	7.6	7.7	7.8	7.9
25	140	142	144	146	148	151	153	155	157	160	162	164	166	169	171	173	175	178
26	145	147	150	152	154	157	159	161	164	166	168	171	173	175	178	180	183	185
27	151	153	156	158	160	163	165	168	170	173	175	177	180	182	185	187	190	192
28	156	159	161	164	166	169	171	174	176	179	181	184	186	189	192	194	197	199
29	162	164	167	170	172	175	177	180	183	185	188	191	193	196	198	201	204	206
30	167	170	173	175	178	181	184	186	189	192	194	197	200	202	205	208	211	213
31	173	176	179	181	184	187	190	193	195	198	201	204	206	209	212	215	218	220
32	179	181	184	187	190	193	196	199	202	204	207	210	213	216	219	222	225	228
33	184	187	190	193	196	199	202	205	208	211	214	217	220	223	226	229	232	235
34	190	193	196	199	202	205	208	211	214	217	220	223	226	229	233	236	239	242
35	195	198	202	205	208	211	214	217	220	224	227	230	233	236	239	243	246	249
36	201	204	207	211	214	217	220	224	227	230	233	237	240	243	246	249	253	256
37	206	210	213	216	220	223	226	230	233	236	240	243	246	250	253	256	260	263
38	212	215	219	222	226	229	233	236	239	243	246	250	253	256	260	263	267	270
39	218	221	225	228	232	235	239	242	246	249	253	256	260	263	267	270	274	277

TABLE A.2 *(continued)*

Strokes/Minute	GALLONS PER STROKE																	
	6.2	6.3	6.4	6.5	6.6	6.7	6.8	6.9	7.0	7.1	7.2	7.3	7.4	7.5	7.6	7.7	7.8	7.9
40	223	227	230	234	238	241	245	248	252	256	259	263	266	270	274	277	281	284
41	229	232	236	240	244	247	251	255	258	262	266	269	273	277	280	284	288	292
42	234	238	242	246	249	253	257	261	265	268	272	276	280	283	287	291	295	299
43	240	244	248	252	255	259	263	267	271	275	279	283	286	290	294	298	302	306
44	246	249	253	257	261	265	269	273	277	281	285	289	293	297	301	305	309	313
45	251	255	259	263	267	271	275	279	283	288	292	296	300	304	308	312	316	320
46	257	261	265	269	273	277	282	286	290	294	298	302	306	310	315	319	323	327
47	262	266	271	275	279	283	288	292	296	300	305	309	313	317	321	326	330	334
48	268	272	276	281	285	289	294	298	302	307	311	315	320	324	328	333	337	341
49	273	278	282	287	291	295	300	304	309	313	318	322	326	331	335	340	344	348
50	279	283	288	292	297	301	306	310	315	319	324	328	333	337	342	346	351	355
51	285	289	294	298	303	308	312	317	321	326	330	335	340	344	349	353	358	363
52	290	295	300	304	309	314	318	323	328	332	337	342	346	351	356	360	365	370
53	296	301	305	310	315	320	324	329	334	339	343	348	353	358	363	367	372	377
54	301	306	311	316	321	326	330	335	340	345	350	355	360	364	369	374	379	384

55	307	312	317	322	327	332	337	342	346	351	356	361	366	371	376	381	386	391
56	312	318	323	328	333	338	343	348	353	358	363	368	373	378	383	388	393	398
57	318	323	328	333	339	344	349	354	359	364	369	374	380	385	390	395	400	405
58	324	329	334	339	345	350	355	360	365	371	376	381	386	391	397	402	407	412
59	329	335	340	345	350	356	361	366	372	377	382	388	393	398	404	409	414	419
60	335	340	346	351	356	362	367	373	378	383	389	394	400	405	410	416	421	427
61	340	346	351	357	362	368	373	379	384	390	395	401	406	412	417	423	428	434
62	346	352	357	363	368	374	379	385	391	396	402	407	413	418	424	430	435	441
63	352	357	363	369	374	380	386	391	397	403	408	414	420	425	431	437	442	448
64	357	363	369	374	380	386	392	397	403	409	415	420	426	432	438	444	449	455
65	363	369	374	380	386	392	398	404	409	415	421	427	433	439	445	450	456	462

TABLE A.3a ANNULAR VELOCITY AROUND THE DRILLPIPE (ft/min)

GPM	Hole Size (in)														
	6-3/4			7-5/8			7-7/8			8-3/8			8-1/2		
	Drill Pipe O.D. (in)														
	3-1/2	4	4-1/2	3-1/2	4	4-1/2	3-1/2	4	4-1/2	3-1/2	4	4-1/2	3-1/2	4	4-1/2
50	37	41	48	–	–	–	–	–	–	–	–	–	–	–	–
60	44	50	58	–	–	–	–	–	–	–	–	–	–	–	–
70	51	58	68	–	–	–	–	–	–	–	–	–	–	–	–
80	59	66	77	–	–	–	–	–	–	–	–	–	–	–	–
90	66	75	87	–	–	–	–	–	–	–	–	–	–	–	–
100	74	83	97	53	58	65	49	53	59	42	45	49	41	44	47
110	81	91	106	59	64	71	54	59	65	47	50	54	45	48	52
120	88	99	116	64	70	78	59	64	70	51	54	59	49	52	57
130	96	108	126	69	76	84	64	69	76	55	59	64	53	57	61
140	103	116	136	75	81	91	69	75	82	59	63	69	57	61	66
150	110	124	145	80	87	97	74	80	88	63	68	74	61	65	71
160	118	133	155	85	93	103	79	85	94	68	72	79	65	70	75
170	125	141	165	91	99	110	84	91	100	72	77	83	69	74	80
180	132	149	174	96	105	116	89	96	106	76	81	88	73	78	85
190	140	157	184	101	110	123	94	101	111	80	86	93	78	83	90
200	147	166	194	107	116	129	98	106	117	85	91	98	82	87	94
210	154	174	203	112	122	136	103	112	123	89	95	103	86	91	99

220	162	182	213	117	128	142	108	117	129	93	100	108	90	96	104
230	169	191	223	123	134	149	113	122	135	97	104	113	94	100	108
240	177	199	232	128	140	155	118	128	141	102	109	118	98	105	113
250	184	207	242	133	145	162	123	133	147	106	113	123	102	109	118
260	191	215	252	139	151	168	128	138	153	110	118	128	106	113	123
270	199	224	261	144	157	175	133	144	158	114	122	133	110	118	127
280	206	232	271	149	163	181	138	149	164	118	127	138	114	122	132
290	213	240	281	155	169	188	143	154	170	123	131	142	118	126	137
300	221	249	290	160	174	194	148	160	176	127	136	147	123	131	141
310	228	257	300	166	180	200	153	165	182	131	140	152	127	135	146
320	235	265	310	171	186	207	158	170	188	135	145	157	131	139	151
330	243	273	319	176	192	213	162	176	194	140	149	162	135	144	155
340	250	282	329	182	198	220	167	181	199	144	154	167	139	148	160
350	257	290	339	187	203	226	172	186	205	148	158	172	143	152	165
360	265	298	348	192	209	233	177	192	211	152	163	177	147	157	170
370	272	307	358	198	215	239	182	197	217	157	167	182	151	161	174
380	279	315	368	203	221	246	187	202	223	161	172	187	155	166	179
390	287	323	377	208	227	252	192	208	229	165	176	192	159	170	184
400	294	332	387	214	233	259	197	213	235	169	181	196	163	174	188
410	302	340	397	219	238	265	202	218	241	174	186	201	167	179	193
420	309	348	407	224	244	272	207	224	246	178	190	206	172	183	198
430	316	356	416	230	250	278	212	229	252	182	195	211	176	187	203
440	324	365	426	235	256	285	217	234	258	186	199	216	180	192	207
450	331	373	436	240	262	291	222	240	264	190	204	221	184	196	212
460	338	381	445	246	267	297	226	245	270	195	208	226	188	200	217
470	346	390	455	251	273	304	231	250	276	199	213	231	192	205	221
480	353	398	465	256	279	310	236	256	282	203	217	236	196	209	226

TABLE A.3a *(continued)*

GPM	Hole Size (in)														
	6-3/4			7-5/8			7-7/8			8-3/8			8-1/2		
	Drill Pipe O.D. (in)														
	3-1/2	4	4-1/2	3-1/2	4	4-1/2	3-1/2	4	4-1/2	3-1/2	4	4-1/2	3-1/2	4	4-1/2
490	360	406	474	262	285	317	241	261	287	207	222	241	200	213	231
500	368	414	484	267	291	323	246	266	293	212	226	246	204	218	236
510	375	423	494	272	296	329	251	272	299	216	231	250	208	222	240
520	382	431	503	277	302	336	256	277	305	220	235	255	212	226	245
530	390	439	513	282	308	342	261	282	311	224	240	260	216	231	250
540	397	448	523	288	313	349	266	288	317	229	244	265	221	235	254
550	405	456	532	293	319	355	271	293	323	233	249	270	225	240	259
560	412	464	542	298	325	362	276	298	328	237	253	275	229	244	264
570	419	472	552	304	331	368	281	303	334	241	258	280	233	248	269
580	427	481	561	309	337	375	286	309	340	245	262	285	237	253	273

TABLE A.3b ANNULAR VELOCITY AROUND THE DRILLCOLLARS (ft/min)

GPM	Hole Size (in)											
	8-1/2						8-5/8					
	Drill Collar Size (in)											
	5-3/4	6	6-1/4	6-1/2	6-3/4	7	5-3/4	6	6-1/4	6-1/2	6-3/4	7
100	63	68	74	82	91	107	60	64	70	77	84	98
110	69	75	82	90	100	117	66	71	77	84	93	108
120	75	82	89	98	109	128	72	77	84	92	101	118
130	82	88	97	106	118	138	78	84	91	100	110	127
140	88	95	104	114	127	149	84	90	98	107	118	137
150	94	102	111	123	136	160	90	97	105	115	127	147
160	101	109	119	131	145	170	96	103	112	122	135	157
170	107	116	126	139	154	181	102	110	119	130	144	167
180	113	123	134	147	163	192	108	116	126	138	152	176
190	119	129	141	155	172	202	114	122	133	145	160	186
200	125	135	148	163	184	211	119	128	139	153	170	193
210	131	142	155	172	193	221	125	134	146	160	179	203
220	138	149	162	180	202	232	130	140	153	168	187	212
230	144	156	170	188	211	242	136	147	160	175	196	222
240	150	162	177	196	220	253	142	153	167	183	204	232
250	156	169	185	204	230	264	148	160	173	191	213	241
260	163	176	192	212	239	274	154	166	180	198	221	251
270	169	183	199	221	248	285	160	172	187	206	230	261
280	175	189	207	229	257	295	166	179	194	214	238	270
290	181	196	214	237	266	306	172	185	201	221	247	280
300	188	203	222	245	276	316	178	192	208	229	255	290
310	194	210	229	253	285	327	184	198	215	236	264	299
320	200	216	236	261	294	337	190	204	222	244	272	309
330	206	223	244	270	303	348	196	211	229	252	281	319
340	213	230	251	278	312	358	202	217	236	259	289	328
350	219	237	259	286	321	369	208	223	243	267	298	338
360	225	243	266	294	331	380	214	230	250	275	306	348
370	231	250	273	302	340	390	219	236	257	282	315	357
380	238	257	281	310	349	401	225	243	264	290	323	367
390	244	264	288	319	358	411	231	249	271	297	332	376
400	250	270	295	327	367	422	237	255	278	305	340	386
410	256	277	303	335	377	432	243	262	284	313	349	396
420	263	284	310	343	386	443	249	268	291	320	357	405
430	269	291	318	351	395	453	255	275	298	328	366	415
440	275	298	325	360	404	464	261	281	305	336	374	425
450	281	304	332	368	413	474	267	287	312	343	383	434
460	288	311	340	376	422	485	273	294	319	351	391	444
470	294	318	347	384	432	496	279	300	326	358	400	454
480	300	325	355	392	441	506	285	306	333	366	408	463
490	306	331	362	400	450	517	291	313	340	374	417	473
500	313	338	369	409	459	527	297	319	347	381	425	483
510	319	345	377	417	468	538	302	326	354	389	434	492
520	325	352	384	425	478	548	308	332	361	397	442	502
530	332	358	391	433	487	559	314	338	368	404	451	512
540	338	365	399	441	496	569	320	345	375	412	459	521
550	344	372	406	449	505	580	326	351	382	419	468	531
560	350	379	414	458	514	590	332	358	389	427	476	541
570	357	385	421	466	524	601	338	364	395	435	485	550
580	363	392	428	474	533	611	344	370	402	442	493	560
590	369	399	436	482	542	622	350	377	409	450	502	570
600	375	406	443	490	551	633	356	383	416	458	510	579
610	382	412	451	498	560	643	362	389	423	465	519	589
620	388	419	458	507	569	654	368	396	430	473	527	599
630	394	426	465	515	579	664	374	402	437	480	536	608

TABLE A.4 DESCRIPTION OF SURFACE EQUIPMENT TYPES

TYPE	STAND PIPE		HOSE		SWIVEL		KELLY	
	LENGTH (ft.)	I.D. (in.)	LENGTH (ft.)	I.D. (in.)	LENGTH (ft.)	I.D. (in.)	LENGTH (ft.)	I.D. (in.)
1	40	3	45	2	4	2	40	2¼
2	40	3½	55	2½	5	2½	40	3¼
3	45	4	55	3	5	2½	40	3¼
4	45	4	55	3	6	3	40	4

TABLE A.5 PRESSURE LOSS THROUGH THE SURFACE EQUIPMENT (psi)

GPM	SURFACE EQUIPMENT TYPE			
	1	2	3	4
50	2	1	—	—
60	3	1	—	—
70	5	1	1	—
80	6	2	1	1
90	8	3	1	1
100	10	3	2	1
110	11	4	2	1
120	14	5	3	2
130	16	5	3	2
140	18	6	4	2
150	21	7	4	3
160	24	8	5	3
170	26	9	5	4
180	29	10	6	4
190	33	11	7	4
200	36	13	8	5
210	39	14	8	6
220	43	15	9	6
230	47	16	10	7
240	50	18	11	7
250	55	19	12	8
260	59	21	13	8
270	63	22	14	9
280	67	24	14	10
290	72	26	15	10
300	77	27	17	11
310	82	29	18	12
320	87	31	19	13
330	92	33	20	13
340	97	35	21	14
350	102	37	22	15
360	108	39	23	16
370	114	41	25	17
380	119	43	26	18
390	125	45	27	18
400	131	47	29	19

GPM	SURFACE EQUIPMENT TYPE			
	1	2	3	4
410	137	49	30	20
420	144	51	31	21
430	150	54	33	22
440	157	56	34	23
450	164	59	36	24
460	170	61	37	25
470	177	64	39	26
480	184	66	40	27
490	192	69	42	28
500	199	71	43	29
510	207	74	45	31
520	214	77	47	32
530	222	80	48	33
540	230	82	50	34
550	238	85	52	35
560	246	88	54	37
570	254	91	56	38
580	262	94	57	39
590	271	97	59	40
600	280	100	61	42
610	288	104	63	43
620	297	107	65	44
630	306	110	67	46
640	315	113	69	47
650	325	117	71	48
660	334	120	73	50
670	343	123	75	51
680	353	127	77	53
690	363	130	79	54
700	373	134	82	56
710	383	137	84	57
720	393	141	86	59
730	403	145	88	60
740	413	148	91	62
750	424	152	93	63
760	434	156	95	65

TABLE A.6 PRESSURE LOSS THROUGH THE DRILLPIPE BORE (psi/1000 ft)

Drill Pipe Size (O.D.) - in	4						4-1/2					
Drill Pipe Weight - lb/ft	14.0				15.7		16.6				20.0	
Type Tool Joint	DBL.S	SH	FH	IF	FH	IF	DBL.S SH	FH	XH	IF	FH XH	IF
GPM												
50	3	2	2	2	3	2	1	1	1	1	1	1
60	4	4	3	3	4	4	2	1	1	1	2	2
70	5	5	5	4	5	5	3	2	2	2	3	3
80	7	7	6	6	7	6	3	3	3	3	4	3
90	9	8	8	7	9	8	4	4	4	3	5	4
100	11	10	9	9	11	10	5	5	4	4	6	6
110	13	12	11	10	13	12	7	6	5	5	7	7
120	16	15	13	12	16	14	8	7	7	6	9	8
130	18	17	15	14	18	17	9	8	8	7	10	9
140	21	20	18	17	21	19	11	9	9	8	12	11
150	24	22	20	19	24	22	12	10	10	10	14	12
160	27	25	23	21	27	25	14	12	11	11	16	14
170	31	28	26	24	31	28	15	13	13	12	17	16
180	34	32	29	27	34	31	17	15	14	14	19	17
190	38	35	32	30	38	34	19	16	16	15	22	19
200	42	38	35	33	42	38	21	18	18	17	24	21
210	46	42	39	36	46	42	23	20	19	18	26	23
220	50	46	42	39	50	45	25	22	21	20	28	26
230	54	50	46	42	54	49	27	24	23	22	31	28
240	58	54	49	46	58	53	30	26	25	24	34	30
250	63	58	53	50	63	58	32	28	27	26	36	33
260	68	63	58	53	68	62	35	30	29	28	39	35
270	73	68	62	57	73	67	37	32	31	30	42	38
280	78	72	66	61	78	71	40	34	33	32	45	40
290	83	77	71	66	83	76	42	37	36	34	48	43

300	89	82	75	70	89	81	45	39	38	36	51	46
310	94	88	80	74	94	86	48	42	40	38	54	49
320	100	93	85	79	100	91	51	44	43	41	58	52
330	106	98	90	84	106	97	54	47	46	43	61	55
340	112	104	95	88	112	102	57	49	48	46	65	58
350	118	110	100	93	118	108	60	52	51	48	68	61
360	125	116	106	98	125	114	64	55	54	51	72	65
370	131	122	111	103	131	120	67	58	56	54	76	68
380	138	128	117	109	138	126	70	61	59	56	80	71
390	145	134	123	114	145	132	74	64	62	59	84	75
400	152	141	129	120	152	139	78	67	65	62	88	79
410	159	148	135	125	159	145	81	70	68	65	92	82
420	167	154	141	131	167	152	85	73	72	68	96	86
430	174	161	147	137	174	159	89	77	75	71	100	90
440	182	168	154	143	182	166	93	80	78	74	105	94
450	189	176	160	149	189	173	97	84	81	77	109	98
460	197	183	167	155	197	180	101	87	85	81	114	102
470	205	190	174	162	205	188	105	91	88	84	118	106
480	214	198	181	168	214	195	109	94	92	87	123	111
490	222	206	188	175	222	203	113	98	96	91	128	115
500	231	214	195	181	231	210	118	102	99	94	133	119
510	239	222	203	188	239	218	122	106	103	98	138	124
520	248	230	210	195	248	226	127	110	107	101	143	128
530	257	238	218	202	257	235	131	114	111	105	148	133
540	266	247	225	209	266	243	136	118	115	109	154	138
550	275	255	233	217	275	251	141	122	119	113	159	143
560	285	264	241	224	285	260	145	126	123	116	164	147
570	294	273	249	232	294	269	150	130	127	120	170	152
580	304	282	258	239	304	278	155	134	131	124	175	157

TABLE A.7 PRESSURE LOSS IN THE DRILLPIPE ANNULUS (psi/1000 ft)

Hole Size (in)	7-5/8						7-7/8				8-3/8			8-1/2		
Drill Pipe Size (in)	3-1/2		4		4-1/2		3-1/2	4	4-1/2		3-1/2	4	4-1/2	3-1/2	4	4-1/2
Tool Joint Type	SH	FH XH IF	FH	IF	FH XH	IF	FH XH IF	FH IF	FH XH	IF	FH XH IF	FH IF	FH XH IF	FH XH IF	FH IF	FH XH IF
GPM																
100	1	1	1	2	2	3	1	1	2	2	1	1	1	–	1	1
110	1	1	2	2	3	3	1	1	2	2	1	1	1	1	1	1
120	1	2	2	2	3	4	1	2	2	3	1	1	2	1	1	1
130	2	2	2	3	4	5	1	2	3	3	1	1	2	1	1	2
140	2	2	3	3	5	5	2	2	3	4	1	1	2	1	1	2
150	2	2	3	4	5	6	2	3	4	4	1	2	2	1	2	2
160	3	3	4	4	6	7	2	3	4	5	1	2	3	1	2	2
170	3	3	4	5	7	8	2	4	5	5	2	2	3	1	2	3
180	3	3	5	5	8	9	3	4	6	6	2	2	3	2	2	3
190	4	4	5	6	9	10	3	4	6	7	2	3	4	2	2	3
200	4	4	6	7	9	11	3	5	7	8	2	3	4	2	3	4
210	4	5	6	7	10	12	4	5	8	8	2	3	5	2	3	4
220	5	5	7	8	11	13	4	6	8	9	3	4	5	2	3	4
230	5	6	8	9	12	14	4	7	9	10	3	4	6	3	4	5
240	6	6	8	9	14	15	5	7	10	11	3	4	6	3	4	5
250	6	7	9	10	15	17	5	8	11	12	3	5	7	3	4	6
260	7	7	10	11	16	18	6	8	12	13	4	5	7	3	5	6
270	7	8	11	12	17	20	6	9	13	14	4	6	8	4	5	7
280	8	8	11	13	18	21	7	10	14	15	4	6	8	4	5	7
290	8	9	12	14	20	23	7	10	15	16	5	6	9	4	6	8
300	9	10	13	15	21	24	8	11	16	17	5	7	10	4	6	8
310	10	10	14	16	23	26	8	12	17	18	5	7	10	5	6	9
320	10	11	15	17	24	27	9	13	18	19	6	8	11	5	7	10
330	11	12	16	18	26	29	9	13	19	21	6	8	11	5	7	10

Security
ROCK BITS & DRILL TOOLS
DRESSER

340	12	12	17	19	27	31	10	14	20	22	6	9	12	6	8	11
350	12	13	18	20	29	33	10	15	21	23	7	9	13	6	8	11
360	13	14	19	21	31	35	11	16	22	24	7	10	14	6	9	12
370	14	14	20	22	32	37	12	17	24	26	8	10	14	7	9	13
380	15	15	21	24	34	39	12	18	25	27	8	11	15	7	10	13
390	15	16	22	25	36	41	13	19	26	29	8	11	16	8	10	14
400	16	17	23	26	38	43	13	20	28	30	9	12	17	8	11	15
410	17	18	25	27	40	45	14	21	29	32	9	13	18	8	11	16
420	18	19	26	29	42	47	15	22	30	33	10	13	19	9	12	16
430	19	20	27	30	44	50	16	23	32	35	10	14	20	9	12	17
440	19	20	28	32	46	52	16	24	33	37	11	15	20	10	13	18
450	20	21	30	33	48	54	17	25	35	38	11	15	21	10	14	19
460	21	22	31	35	50	57	18	26	37	40	12	16	22	11	14	20
470	22	23	32	36	52	59	19	27	38	42	12	17	23	11	15	21
480	23	24	34	38	54	62	19	28	40	44	13	17	24	12	16	21
490	24	25	35	39	57	64	20	30	41	45	13	18	25	12	16	22
500	25	26	37	41	59	67	21	31	43	47	14	19	26	12	17	23
510	26	28	38	42	61	70	22	32	45	49	14	20	27	13	18	24
520	27	29	40	44	64	72	23	33	47	51	15	20	29	14	18	25
530	28	30	41	46	66	75	24	35	48	53	16	21	30	14	19	26
540	29	31	43	48	69	78	25	36	50	55	16	22	31	15	20	27
550	30	32	44	49	71	81	25	37	52	57	17	23	32	15	20	28
560	32	33	46	51	74	84	26	39	54	59	17	24	33	16	21	29
570	33	34	47	53	77	87	27	40	56	61	18	25	34	16	22	30
580	34	36	49	55	79	90	28	42	58	64	19	25	36	17	23	31
590	35	37	51	57	82	93	29	43	60	66	19	26	37	17	24	32
600	36	38	53	59	85	96	30	45	62	68	20	27	38	18	24	33
610	–	–	–	–	–	–	31	46	64	70	21	28	39	19	25	35
620	–	–	–	–	–	–	32	48	66	73	21	29	41	19	26	36
630	–	–	–	–	–	–	33	49	68	75	22	30	42	20	27	37

TABLE A.8 PRESSURE LOSS THROUGH THE DRILLCOLLAR BORE (psi/100 ft)

GPM	DRILL COLLAR BORE (in)								
	1	1-1/4	1-1/2	1-3/4	2	2-1/4	2-1/2	2-13/16	3
50	88	29	12	5	3	1	1	–	–
60	123	41	17	8	4	2	1	–	–
70	164	55	22	10	5	3	1	1	–
80	211	71	29	13	7	4	2	1	1
90	263	88	36	17	9	5	3	1	1
100	320	108	44	21	11	6	3	2	1
110	382	129	53	25	13	7	4	2	1
120	449	151	62	29	15	8	5	2	2
130	521	176	72	34	17	10	6	3	2
140	598	202	83	39	20	11	6	3	2
150	680	230	94	44	23	13	7	4	3
160	767	259	106	50	26	14	8	5	3
170	858	290	119	56	29	16	9	5	4
180	955	322	133	62	32	18	11	6	4
190	–	357	147	69	36	20	12	6	5
200	–	392	161	76	40	22	13	7	5
210	–	430	177	83	43	24	14	8	6
220	–	469	193	91	47	26	16	9	6
230	–	509	210	99	51	29	17	9	7
240	–	551	227	107	56	31	18	10	7
250	–	594	245	115	60	34	20	11	8
260	–	640	263	124	65	36	22	12	9
270	–	686	283	133	69	39	23	13	9
280	–	734	302	143	74	42	25	14	10
290	–	784	323	152	79	45	27	15	11
300	–	835	344	162	85	47	28	16	11
310	–	887	365	173	90	51	30	17	12
320	–	941	388	183	95	54	32	18	13
330	–	997	411	194	101	57	34	19	14
340	–	–	434	205	107	60	36	20	14
350	–	–	458	216	113	63	38	21	15
360	–	–	483	228	119	67	40	22	16
370	–	–	508	240	125	70	42	23	17
380	–	–	534	252	132	74	44	25	18
390	–	–	560	265	138	78	46	26	19
400	–	–	587	277	145	81	49	27	20
410	–	–	615	291	152	85	51	29	21
420	–	–	643	304	159	89	53	30	22
430	–	–	672	317	166	93	56	31	23
440	–	–	702	331	173	97	58	33	24
450	–	–	731	346	180	102	61	34	25
460	–	–	762	360	188	106	63	35	26

TABLE A.9 PRESSURE LOSS IN THE DRILLCOLLAR ANNULUS (psi/100 ft)

GPM	Hole Size (in)											
	8-1/2						8-5/8					
	Drill Collar Size (in)											
	5-3/4	6	6-1/4	6-1/2	6-3/4	7	5-3/4	6	6-1/4	6-1/2	6-3/4	7
200	1	1	1	2	3	4	1	1	1	2	2	3
210	1	1	2	2	3	5	1	1	1	2	2	4
220	1	1	2	2	3	5	1	1	1	2	3	4
230	1	1	2	3	4	6	1	1	2	2	3	4
240	1	2	2	3	4	6	1	1	2	2	3	5
250	1	2	2	3	4	7	1	1	2	2	3	5
260	1	2	2	3	5	7	1	2	2	3	4	6
270	1	2	3	3	5	8	1	2	2	3	4	6
280	2	2	3	4	5	8	1	2	2	3	4	6
290	2	2	3	4	6	9	1	2	2	3	5	7
300	2	2	3	4	6	10	2	2	3	4	5	7
310	2	3	3	5	7	10	2	2	3	4	5	8
320	2	3	4	5	7	11	2	2	3	4	6	8
330	2	3	4	5	8	12	2	2	3	4	6	9
340	2	3	4	6	8	12	2	3	3	5	6	10
350	2	3	4	6	8	13	2	3	4	5	7	10
360	3	3	5	6	9	14	2	3	4	5	7	11
370	3	4	5	7	9	15	2	3	4	5	8	11
380	3	4	5	7	10	15	3	3	4	6	8	12
390	3	4	5	7	11	16	3	3	4	6	8	13
400	3	4	6	8	11	17	3	4	5	6	9	13
410	3	4	6	8	12	18	3	4	5	7	9	14
420	4	5	6	8	12	19	3	4	5	7	10	14
430	4	5	6	9	13	20	3	4	5	7	10	15
440	4	5	7	9	13	21	3	4	6	8	11	16
450	4	5	7	10	14	22	4	5	6	8	11	17
460	4	6	7	10	15	22	4	5	6	8	12	17
470	5	6	8	11	15	23	4	5	6	9	12	18
480	5	6	8	11	16	24	4	5	7	9	13	19
490	5	6	8	11	17	25	4	5	7	9	13	20
500	5	7	9	12	17	27	4	6	7	10	14	21
510	5	7	9	12	18	28	5	6	8	10	14	21
520	6	7	9	13	19	29	5	6	8	11	15	22
530	6	7	10	13	19	30	5	6	8	11	16	23
540	6	8	10	14	20	31	5	6	8	11	16	24
550	6	8	11	14	21	32	5	7	9	12	17	25

TABLE A.10 PRESSURE LOSS THROUGH THE JET NOZZLES (psi)

GPM	JET NOZZLE SIZE/AREA*											
	7-7-7 0.1127	7-7-8 0.1242	7-8-8 0.1358	8-8-8 0.1473	8-8-9 0.1603	8-9-9 0.1733	9-9-9 0.1864	9-9-10 0.2009	9-10-10 0.2155	10-10-10 0.2301	10-10-11 0.2462	10-11-11 0.2623
50	181	149	125	106	90	77	66	57	50	43	38	33
60	261	215	180	153	129	110	95	82	71	63	55	48
70	355	292	245	208	176	150	130	112	97	85	74	66
80	464	382	320	272	229	196	170	146	127	111	97	86
90	587	483	405	344	290	248	215	185	161	141	123	108
100	725	597	500	425	358	307	265	228	198	174	152	134
110	877	722	605	514	434	371	321	276	240	210	184	162
120	1043	859	720	612	516	441	382	328	286	250	219	193
130	1224	1008	845	718	606	518	448	385	335	294	257	226
140	1420	1169	979	832	702	601	520	447	389	341	298	262
150	1630	1342	1124	956	806	690	597	513	446	391	342	301
160	1855	1527	1279	1087	918	785	679	584	508	445	389	343
170	2094	1724	1444	1227	1036	886	766	659	573	503	439	387
180	2347	1933	1619	1376	1161	993	859	739	642	564	492	434
190	2615	2154	1804	1533	1294	1107	957	823	716	628	548	483
200	2898	2386	1999	1699	1434	1226	1061	912	793	696	608	535
210	–	2631	2204	1873	1581	1352	1169	1006	874	767	670	590
220	–	2887	2419	2055	1735	1484	1283	1104	960	842	735	648
230	–	–	2644	2247	1896	1621	1403	1206	1049	920	804	708
240	–	–	2878	2446	2064	1766	1527	1314	1142	1002	875	771

250	–	–	–	2654	2240	1916	1657	1425	1239	1087	950	837
260	–	–	–	2871	2423	2072	1792	1542	1340	1176	1027	905
270	–	–	–	–	2613	2235	1933	1663	1445	1268	1108	976
280	–	–	–	–	2810	2403	2079	1788	1554	1364	1191	1049
290	–	–	–	–	–	2578	2230	1918	1667	1463	1278	1126
300	–	–	–	–	–	2759	2386	2053	1784	1566	1367	1205
310	–	–	–	–	–	2946	2548	2192	1905	1672	1460	1286
320	–	–	–	–	–	–	2715	2335	2030	1781	1556	1371
330	–	–	–	–	–	–	2887	2484	2159	1894	1655	1458
340	–	–	–	–	–	–	–	2637	2292	2011	1756	1547
350	–	–	–	–	–	–	–	2794	2429	2131	1861	1640
360	–	–	–	–	–	–	–	2956	2570	2254	1969	1735
370	–	–	–	–	–	–	–	–	2714	2381	2080	1832
380	–	–	–	–	–	–	–	–	2863	2512	2194	1933
390	–	–	–	–	–	–	–	–	–	2646	2311	2036
400	–	–	–	–	–	–	–	–	–	2783	2431	2142
410	–	–	–	–	–	–	–	–	–	2924	2554	2250
420	–	–	–	–	–	–	–	–	–	–	2680	2361
430	–	–	–	–	–	–	–	–	–	–	2809	2475
440	–	–	–	–	–	–	–	–	–	–	2941	2591
450	–	–	–	–	–	–	–	–	–	–	–	2710
460	–	–	–	–	–	–	–	–	–	–	–	2832
470	–	–	–	–	–	–	–	–	–	–	–	2957

*Nozzle Size (32nd in.)   Nozzle Area (Sq. in.)

TABLE A.11 JET VELOCITY (ft/sec)

GPM	JET NOZZLE SIZE/AREA*											
	7-7-7 0.1127	7-7-8 0.1242	7-8-8 0.1358	8-8-8 0.1473	8-8-9 0.1603	8-9-9 0.1733	9-9-9 0.1864	9-9-10 0.2009	9-10-10 0.2155	10-10-10 0.2301	10-10-11 0.2462	10-11-11 0.2623
50	142	129	118	109	100	92	86	80	74	70	65	61
60	170	155	141	130	120	111	103	96	89	83	78	73
70	199	180	165	152	140	129	120	111	104	97	91	85
80	227	206	189	174	160	148	137	127	119	111	104	98
90	255	232	212	196	180	166	155	143	134	125	117	110
100	284	258	236	217	200	185	172	159	148	139	130	122
110	312	283	259	239	220	203	189	175	163	153	143	134
120	341	309	283	261	240	222	206	191	178	167	156	146
130	369	335	306	282	260	240	223	207	193	181	169	159
140	397	361	330	304	279	258	240	223	208	195	182	171
150	426	386	354	326	299	277	258	239	223	209	195	183
160	454	412	377	348	319	295	275	255	238	223	208	195
170	482	438	401	369	339	314	292	271	252	236	221	207
180	511	464	424	391	359	332	309	287	267	250	234	220
190	539	489	448	413	379	351	326	303	282	264	247	232
200	568	515	471	435	399	369	343	318	297	278	260	244
210	596	541	495	456	419	388	361	334	312	292	273	256
220	624	567	519	478	439	406	378	350	327	306	286	268
230	653	592	542	500	459	425	395	366	341	320	299	281

240	681	618	566	522	479	443	412	382	356	334	312	293
250	710	644	589	543	499	462	429	398	371	348	325	305
260	738	670	613	565	519	480	446	414	386	362	338	317
270	766	695	636	587	539	498	464	430	401	375	351	329
280	795	721	660	608	559	517	481	446	416	389	364	342
290	–	–	–	630	579	535	498	462	431	403	377	354
300	–	–	–	652	599	554	515	478	445	417	390	366
310	–	–	–	–	–	–	532	494	460	431	403	378
320	–	–	–	–	–	–	549	510	475	445	416	390
330	–	–	–	–	–	–	567	525	490	459	429	403
340	–	–	–	–	–	–	584	541	505	473	442	415
350	–	–	–	–	–	–	–	–	520	487	455	427
360	–	–	–	–	–	–	–	–	535	501	468	439
370	–	–	–	–	–	–	–	–	549	515	481	451
380	–	–	–	–	–	–	–	–	564	528	494	464
390	–	–	–	–	–	–	–	–	–	–	507	476
400	–	–	–	–	–	–	–	–	–	–	520	488
410	–	–	–	–	–	–	–	–	–	–	533	500
420	–	–	–	–	–	–	–	–	–	–	–	512
430	–	–	–	–	–	–	–	–	–	–	–	525
440	–	–	–	–	–	–	–	–	–	–	–	537
450	–	–	–	–	–	–	–	–	–	–	–	549
460	–	–	–	–	–	–	–	–	–	–	–	561

*Nozzle Size (32nd in.) Nozzle Area (Sq. in.)

Appendix B

The following is a reprint of a formation testing service report (Well #2, DST #2). This report is reprinted courtesy of Halliburton Services, Duncan, Oklahoma.

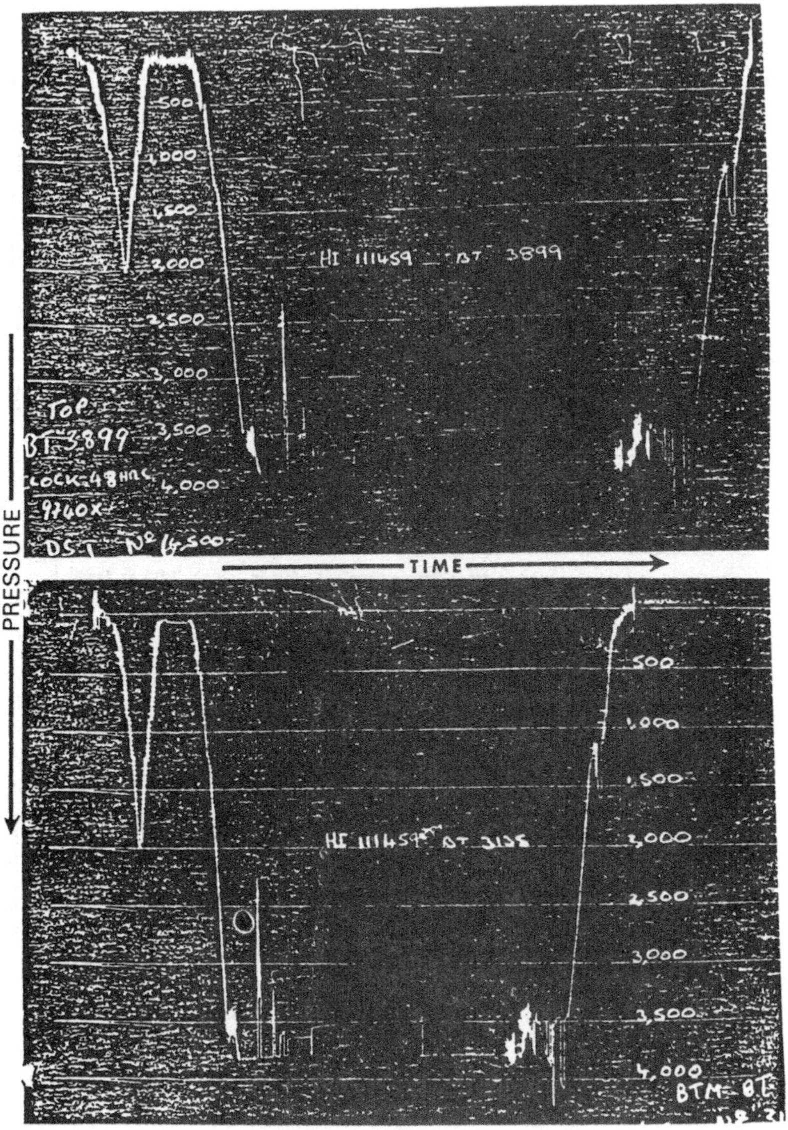

Each Horizontal Line Equal to 1000 psi

FLUID SAMPLE DATA	
Sampler Pressure ____ P.S.I.G. at Surface	Date 6th May, 1976. — Ticket Number HI 111459
Recovery: Cu. Ft. Gas ____	Kind of Job 7" Casing test — Halliburton District
cc. Oil ____	Tester — Witness
cc. Water ____	Drilling Contractor International
cc. Mud ____	**EQUIPMENT & HOLE DATA**
Tot. Liquid cc. ____	Formation Tested
Gravity ____ ° API @ ____ °F.	Elevation ____ Ft.
Gas/Oil Ratio ____ cu. ft./bbl.	Net Productive Interval 142' Ft.
RESISTIVITY / CHLORIDE CONTENT	All Depths Measured From K.B.
Recovery Water ____ @ ____ °F. ____ ppm	Total Depth 8685.19' Ft.
Recovery Mud ____ @ ____ °F. ____ ppm	Main Hole/Casing Size 7"
Recovery Mud Filtrate ____ @ ____ °F. ____ ppm	Drill Collar Length 560.20' I.D. 2.25"
Mud Pit Sample ____ @ ____ °F. ____ ppm	Drill Pipe Length ____ I.D.
Mud Pit Sample Filtrate ____ @ ____ °F. ____ ppm	Packer Depth(s) 8640.42' Ft.
Mud Weight 8.5 vis cp	Depth Tester Valve 8431.17' Ft.

Cushion TYPE H2O AMOUNT 1000'. Ft. Depth Back Pres. Valve — Surface Choke — Bottom Choke 2.25"

Recovered ____ Feet of

Recovered ____ Feet of

Recovered ____ Feet of

Recovered ____ Feet of

Recovered ____ Feet of

Meas. from Tester Valve

Remarks R.I.H. roughly ½ way when got a leak - Came out of hole losing fluid as fast as pulled - Pulled to ful-flo hydrospring and found ball open - Collet connector was not made up to mandrel - Redressed tool R.I.H. - Dropped bar to reverse out, did not reach bottom found it on top of top slip joint - On way out, well flowing had to kill it three times. *No reading possible

TEMPERATURE	Gauge No. 3899 Depth: 8675.20" Ft.		Gauge No. 3135 Depth: 8681.03 Ft.		Gauge No. Depth: Ft.		TIME	
Est. 290 °F.	48 Hour Clock / Blanked Off Yes		72 Hour Clock / Blanked Off Yes		Hour Clock / Blanked Off		Tool Opened A.M. P.M.	
Actual °F.	Pressures		Pressures		Pressures		Opened Bypass A.M. P.M.	
	Field	Office	Field	Office	Field	Office	Reported Minutes	Computed Minutes
Initial Hydrostatic	3848.4	3836	3008.3	3821				
First Period Flow Initial	2323.6	2327	2260.9	2269			—	—
First Period Flow Final	3142.6	*	3150.5	3141			7	11
First Period Closed in	3822.4	3830	3808.3	3811			68	68
Second Period Flow Initial	3234.2	3257	3223.6	3220			—	—
Second Period Flow Final	3482.9	3494	3428.2	3453			783	783
Second Period Closed in	3809.3	*	3793.7	*			476	478
Third Period Flow Initial							—	—
Third Period Flow Final								
Third Period Closed in								
Final Hydrostatic	3874.5	3813	3837.6	3786			—	—

Legal Location Sec. - Twp. - Rng.

Lease Name

2. Well No.

6. Test No.

8678'-8825' + 8931'-8983' Tested Interval

Field Area Offshore

County

State

Oil Co. Lease Owner/Company Name

Casing perfs 124' Bottom choke 0.75" Surf. temp 75 °F Ticket No. HI 111459

Gas gravity 0.957 Oil gravity 32.0 GOR 40

Spec. gravity ______ Chlorides ______ ppm Res. ______ @ ______ °F

INDICATE TYPE AND SIZE OF GAS MEASURING DEVICE USED ______

Date 6/5/76 Time a.m. p.m.	Choke Size	Surface Pressure psi	Gas Rate MCF	Liquid Rate BPD	Remarks
07.40					Start in hole with D.S.T. 6.
10.45					Start out of hole - Leaking ball-valve.
12.50					Out of hole.
15.15					R.I.H. Again - Ball valve O.K.
20.20					Test manifold to 3000 PSI.
20.54					Packer Set.
21.12					Tool opened - 18 mins to open.
21.19					Ratchet tool closed - 7 min flow.
22.15					Ratchet tool open - 13 mins to open.
22.28					Tool open - Strong blow.
22.36	16/64	Adjustable			Cushion to surface.
22.39					Mud to surface.
22.45					Oil/Mud to surface.
23.00	32/64	524		Temp 65 F	Choke change. Trace to H2S.
23.05	64/64	340		" " "	
23.24	32/64				Choke change.
23.25	"	660		" 65 F	
24.00	"	365		" " "	
01.45		415		" 72 F	
01.49	80/64		Choke change	- Gradual	increased to maxium flo-rate.
01.50	"	227		Temp 76 F	
02.56	112/64	227		" 86 F	Gradual opening to 128/64"
03.29	128/64	230		" 88 F	
07.30	"	242		" 105 F	6227 bbs/day.
09.30	"	242		" 105 F	6070 bbs/day.
11.31					Ratchet tool closed.

Casing perfs.______ Bottom choke______ Surf. temp.______ °F Ticket No. HI 111459
Gas gravity______ Oil gravity______ GOR______
Spec. gravity______ Chlorides______ ppm Res.______ @______ °F

INDICATE TYPE AND SIZE OF GAS MEASURING DEVICE USED______

Date Time a.m. p.m.	Choke Size	Surface Pressure psi	Gas Rate MCF	Liquid Rate BPD	Remarks
14.15					Shear pump out - Rev. Circ. out
19.27					Pull packer - Well started flowing -
					Stopped pulling and tried to pump into
					formation. Pulled out of 7" - Killed
					well - Twice more on trip out of hole
					Trace H_2 S All way through.
05.55					B.T.'s out of hole

TICKET NO. BI 111459

DESCRIPTION.	O.D.	I.D.	LENGTH.	DEPTH.
5" D.P. & SS.T.T.			5846.52" ft.	
X/O 3½" IF PIN x 4½" IF BOX	5.00"	4.27"	18.00"	
8 STDS 3½" IF D.P.	3.75"	2.76"	757.36 ft.	
X/O 2½ IF PIN 3½ IF BOX	4.75"	2.12"	8.12"	
1 x SLIP JOINT	4.38"	2.00"	179.50"	
X/O 3½" IF PIN x 2½" IF BOX.	4.75"	2.37"	8.25"	
15 STDS 3½" IF D.P.	3.50"	2.76"	1407.95 ft.	
7 JOINTS 4½" D.C.'s	4.75"	2.25"	2481.24"	
PUMP OUT SUB	5.00"	2.25"	12.00"	
1 STD 4½" D.C.'s	4.75"	2.25"	1060.56"	
3½" BAR DROP SUB	5.00"	2.25"	12.00"	
1 STD 4½" D.C.'s	4.75"	2.25"	1071.24"	
X/O 3½ FH PIN x 3½ IF BOX	4.75"	2.25"	12.00"	
5" HANDLING SUB	5.00"	2.50"	54.00"	
5" FUL-FLO HYDROSPRING	5.00"	2.25"	166.68"	8431.17"
X/O 3½ IF PIN x 3½" FHB	4.75"	2.25"	12.00"	
1 STD 4½" D.C.'s	4.75"	2.25"	1057.8"	
X/O 2½" IF PIN x 3½" IF BOX	6.75"	1.50"	14.25"	
1 x SLIP JOINT	4.38"	2.00"	179.50"	
X/O 3½ IF PIN x 2½" IF BOX	6.75"	2.25"	8.00" (87.74')	
1 STD 4½" D.C.'s	4.75"	2.25"	1052.80"	
X/O 2½" EUE P x 3½ IF BOX	4.75"	2.25"	11.25	
7" CIRCULATING VALVE	4.62"	2.44"	35.50"	
7" SAFETY JOINT	5.00"	2.44"	27.50" (32.00" BELOW)	
PACKER 7" R.T.T.S.	5.75"	2.18"	61.00"	8640.42 ft.
X/O 2 3/8" IF PIN x 2½" EUE. P	3.37"	1.75"	8.50"	
6 JOINTS ANCHOR PIPE	3.75"	1.75"	360.00"	
B.T. 500 TEMP RECORDER	3.75"		16.87"	
BLANKED OFF B.T. CASE	3.75"		70.00"	8675.20' ft.
BLANKED OFF B.T. CASE	3.75"		50.00"	8681.03' ft.
TOTAL DEPTH				8685.19 Ft.

COMPANY: OIL CO.
TICKET NO: H111459
GAUGE NO: 3899

FIRST FLOW 1, 5 MINUTE INTERVALS. LAST INTERVAL= 6 MINS.
TOTAL 11 MINUTES

	TIME		PRESSURE		
	MINUTES	INCHES DEFL.	INCHES DEFL.	PSIG	DELTA P
P0	0	.000	1.743	2327	0
P1	5	.008	NO READINGS POSSIBLE		
P2	11	.018			

COMPANY: OIL CO.
TICKET NO: H111459
GAUGE NO: 3899

FIRST CLOSURE 13, 5 MINUTE INTERVALS. LAST INTERVAL= 3 MINS.
TOTAL 68 MINUTES

	TIME		PRESSURE		
	MINUTES	INCHES DEFL.	INCHES DEFL.	PSIG	DELTA P
P0	0	.000	NO READINGS POSSIBLE		
P1	5	.008	2.887	3831	0
P2	10	.017	2.887	3831	0
P3	15	.025	2.887	3831	0
P4	20	.034	2.887	3831	0
P5	25	.042	2.887	3831	0
P6	30	.051	2.887	3831	0
P7	35	.059	2.887	3831	0
P8	40	.067	2.886	3830	1
P9	45	.076	2.886	3830	0
P10	50	.084	2.886	3830	0
P11	55	.093	2.886	3830	0
P12	60	.101	2.886	3830	0
P13	65	.110	2.886	3830	0
P14	68	.114	2.886	3830	0

SECOND FLOW

77, 10 MINUTE INTERVALS. LAST INTERVAL= 13 MINS.
TOTAL 783 MINUTES

	TIME		PRESSURE		
	MINUTES	INCHES DEFL.	INCHES DEFL.	PSIG	DELTA P
P0	0	.000	2.448	3257	0
P1	10	.017	2.868	3806	+549
P2	20	.034	2.883	3826	+ 20
P3	30	.051	2.882	3825	−1
P4	40	.067	2.791	3706	−119
P5	50	.084	2.729	3625	− 81
P6	60	.101	2.879	3821	+196
P7	70	.118	2.882	3825	+ 4
P8	80	.135	2.774	3684	−141
P9	90	.152	2.771	3680	− 4
P10	100	.169	2.770	3678	− 2
P11	110	.186	2.767	3674	− 4
P12	120	.202	2.765	3672	− 2
P13	130	.219	2.763	3669	− 3
P14	140	.236	2.764	3671	+2
P15	150	.253	2.769	3677	+6
P16	160	.270	2.765	3672	−5
P17	170	.287	2.764	3671	−1
P18	180	.304	2.762	3668	− 3
P19	190	.321	2.761	3667	− 1
P20	200	.337	2.760	3665	− 2
P21	210	.354	2.703	3591	− 74
P22	220	.371	2.701	3588	− 3
P23	230	.388	2.699	3586	− 2
P24	240	.405	2.697	3583	− 3
P25	250	.422	2.695	3580	− 3
P26	260	.439	2.693	3578	− 2
P27	270	.456	2.691	3575	− 3
P28	280	.472	2.690	3574	− 1
P29	290	.489	2.689	3573	− 1
P30	300	.506	2.689	3573	0
P31	310	.523	2.686	3569	− 4
P32	320	.540	2.687	3570	+ 1
P33	330	.557	2.686	3569	− 1
P34	340	.574	2.684	3566	− 3
P35	350	.591	2.683	3565	− 1
P36	360	.607	2.683	3565	0
P37	370	.624	2.685	3567	+ 2
P38	380	.641	2.685	3567	0
P39	390	.658	2.679	3560	− 7
P40	400	.675	2.678	3558	− 2
P41	410	.692	2.674	3553	− 5
P42	420	.709	2.672	3550	− 3

COMPANY: OIL CO.
TICKET NO: H111459
GAUGE NO: 3899

SECOND FLOW CONTINUED

	TIME		PRESSURE		
	MINUTES	INCHES DEFL.	INCHES DEFL.	PSIG	DELTA P
P43	430	.726	2.671	3549	− 1
P44	440	.742	2.671	3549	0
P45	450	.759	2.670	3548	− 1
P46	460	.776	2.669	3546	− 2
P47	470	.793	2.669	3546	0
P48	480	.810	2.668	3545	− 1
P49	490	.827	2.668	3545	0
P50	500	.844	2.668	3545	0
P51	510	.861	2.671	3549	+ 4
P52	520	.877	2.675	3554	+ 5
P53	530	.894	2.675	3554	0
P54	540	.911	2.674	3553	− 1
P55	550	.928	2.673	3552	− 1
P56	560	.945	2.670	3548	− 4
P57	570	.962	2.669	3546	− 2
P58	580	.979	2.670	3548	+ 2
P59	590	.996	2.670	3548	0
P60	600	1.012	2.669	3546	− 2
P61	610	1.029	2.667	3544	− 2
P62	620	1.046	2.667	3544	0
P63	630	1.063	2.667	3544	0
P64	640	1.080	2.667	3544	0
P65	650	1.097	2.667	3544	0
P66	660	1.114	2.667	3544	0
P67	670	1.131	2.667	3544	0
P68	680	1.147	2.667	3544	0
P69	690	1.164	2.667	3544	0
P70	700	1.181	2.667	3544	0
P71	710	1.198	2.667	3544	0
P72	720	1.215	2.667	3544	0
P73	730	1.232	2.667	3544	0
P74	740	1.249	2.666	3543	− 1
P75	750	1.266	2.666	3543	0
P76	760	1.282	2.666	3543	0
P77	770	1.299	2.666	3543	0
P78	783	1.321	2.629	3494	− 49

COMPANY: OIL CO.
TICKET NO: H111459
GAUGE NO: 3899

SECOND CLOSURE 47, 10 MINUTE INTERVALS. LAST INTERVAL= 8 MINS.
TOTAL 478 MINUTES

	TIME		PRESSURE		
	MINUTES	INCHES DEFL.	INCHES DEFL.	PSIG	DELTA P
P0	0	.000	2.629	3494	0
P1	10	.017	2.881	3823	+329
P2	20	.034	2.881	3823	0
P3	30	.051	2.881	3823	0
P4	40	.067	2.882	3825	+2
P5	50	.084	2.882	3825	0
P6	60	.101	2.882	3825	0
P7	70	.118	2.882	3825	0
P8	80	.135	2.882	3825	0
P9	90	.152	2.882	3825	0
P10	100	.169	2.882	3825	0
P11	110	.186	2.882	3825	0
P12	120	.202	2.882	3825	0
P13	130	.219	2.882	3825	0
P14	140	.236	2.881	3823	−2
P15	150	.253	2.881	3823	0
P16	160	.270	2.881	3823	0
P17	170	.287	2.881	3823	0
P18	180	.304	2.881	3823	0
P19	190	.321	2.882	3825	+2
P20	200	.337	2.882	3825	0
P21	210	.354	2.882	3825	0
P22	220	.371	2.882	3825	0
P23	230	.388	2.882	3825	0
P24	240	.405	2.882	3825	0
P25	250	.422	2.882	3825	0
P26	260	.439	2.882	3825	0
P27	270	.456	2.882	3825	0
P28	280	.472	2.882	3825	0
P29	290	.489	2.882	3825	0
P30	300	.506	2.881	3823	−2
P31	310	.523	2.881	3823	0
P32	320	.540	2.881	3823	0
P33	330	.557	2.881	3823	0
P34	340	.574	2.882	3825	+2
P35	350	.591	2.882	3825	0
P36	360	.607	2.882	3825	0
P37	370	.624	2.882	3825	0
P38	380	.641	2.882	3825	0
P39	390	.658	2.882	3825	0
P40	400	.675	2.882	3825	0
P41	410	.692	2.882	3825	0
P42	420	.709	2.882	3825	0

SECOND CLOSURE CONTINUED

	TIME		PRESSURE		
	MINUTES	INCHES DEFL.	INCHES DEFL.	PSIG	DELTA P
P43	430	.726	2.882	3825	0
P44	440	.742	2.882	3825	0
P45	450	.759	2.882	3825	0
P46	460	.776	NO READINGS POSSIBLE		
P47	470	.793			
P48	478	.806			

COMPANY: OIL CO.
TICKET NO: H111459
GAUGE NO: 3135

FIRST FLOW 1, 5 MINUTE INTERVALS. LAST INTERVAL= 6 MINS.
TOTAL 11 MINUTES

	TIME		PRESSURE		
	MINUTES	INCHES DEFL.	INCHES DEFL.	PSIG	DELTA P
P0	0	.000	1.566	2269	0
P1	5	.006	1.996	2896	627
P2	11	.012	2.164	3141	245

COMPANY: OIL CO.
TICKET NO: H111459
GAUGE NO: 3135

FIRST CLOSURE 13, 5 MINUTE INTERVALS. LAST INTERVAL= 3 MINS.
TOTAL 68 MINUTES

	TIME		PRESSURE		
	MINUTES	INCHES DEFL.	INCHES DEFL.	PSIG	DELTA P
P0	0	.000	2.164	3141	0
P1	5	.006	2.625	3815	+ 674
P2	10	.011	2.625	3815	0
P3	15	.017	2.624	3814	1
P4	20	.022	2.624	3814	0
P5	25	.028	2.624	3814	0
P6	30	.033	2.624	3814	0
P7	35	.039	2.623	3812	− 2
P8	40	.045	2.623	3812	0
P9	45	.050	2.623	3812	0
P10	50	.056	2.623	3812	0
P11	55	.061	2.623	3812	0
P12	60	.067	2.623	3812	0
P13	65	.072	2.622	3811	− 1
P14	68	.076	2.622	3811	0

COMPANY: OIL CO.
TICKET NO: HI11459
GAUGE NO: 3135

SECOND FLOW 77, 10 MINUTE INTERVALS. LAST INTERVAL= 13 MINS.
TOTAL 783 MINUTES

	TIME		PRESSURE		
	MINUTES	INCHES DEFL.	INCHES DEFL.	PSIG	DELTA P
P0	0	.000	2.218	3220	0
P1	10	.011	2.597	3774	554
P2	20	.022	2.619	3807	33
P3	30	.033	2.619	3807	0
P4	40	.045	2.493	3622	185
P5	50	.056	2.476	3597	25
P6	60	.067	2.614	3799	202
P7	70	.078	2.615	3801	2
P8	80	.089	2.518	3659	142
P9	90	.100	2.514	3653	6
P10	100	.111	2.513	3652	1
P11	110	.123	2.510	3647	5
P12	120	.134	2.507	3643	4
P13	130	.145	2.506	3641	2
P14	140	.156	2.505	3640	1
P15	150	.167	2.509	3646	6
P16	160	.178	2.507	3643	3
P17	170	.189	2.506	3641	2
P18	180	.201	2.504	3638	3
P19	190	.212	2.503	3637	1
P20	200	.223	2.502	3635	2
P21	210	.234	2.450	3559	76
P22	220	.245	2.447	3555	4
P23	230	.256	2.445	3552	3
P24	240	.267	2.443	3549	3
P25	250	.278	2.440	3545	4
P26	260	.290	2.439	3543	2
P27	270	.301	2.438	3542	1
P28	280	.312	2.436	3539	3
P29	290	.323	2.435	3538	1
P30	300	.334	2.435	3538	0
P31	310	.345	2.435	3538	0
P32	320	.356	2.432	3533	5
P33	330	.368	2.431	3532	1
P34	340	.379	2.431	3532	0
P35	350	.390	2.430	3530	2
P36	360	.401	2.429	3529	1
P37	370	.412	2.430	3530	1
P38	380	.423	2.430	3530	0
P39	390	.434	2.425	3523	7
P40	400	.446	2.424	3521	2
P41	410	.457	2.422	3519	2
P42	420	.468	2.418	3513	6

COMPANY: OIL CO.
TICKET NO: H111459
GAUGE NO: 3135

SECOND FLOW CONTINUED

	TIME		PRESSURE		
	MINUTES	INCHES DEFL.	INCHES DEFL.	PSIG	DELTA P
P43	430	.479	2.417	3511	2
P44	440	.490	2.416	3510	1
P45	450	.501	2.416	3510	0
P46	460	.512	2.415	3508	2
P47	470	.524	2.413	3505	3
P48	480	.535	2.413	3505	0
P49	490	.546	2.413	3505	0
P50	500	.557	2.412	3504	1
P51	510	.568	2.413	3505	1
P52	520	.579	2.418	3513	8
P53	530	.590	2.418	3513	0
P54	540	.602	2.418	3513	0
P55	550	.613	2.418	3513	0
P56	560	.624	2.416	3510	3
P57	570	.635	2.414	3507	3
P58	580	.646	2.414	3507	0
P59	590	.657	2.414	3507	0
P60	600	.668	2.415	3508	1
P61	610	.680	2.412	3504	4
P62	620	.691	2.412	3504	0
P63	630	.702	2.412	3504	0
P64	640	.713	2.412	3504	0
P65	650	.724	2.412	3504	0
P66	660	.735	2.412	3504	0
P67	670	.746	2.412	3504	0
P68	680	.758	2.412	3504	0
P69	690	.769	2.411	3502	2
P70	700	.780	2.411	3502	0
P71	710	.791	2.411	3502	0
P72	720	.802	2.411	3502	0
P73	730	.813	2.411	3502	0
P74	740	.824	2.410	3501	1
P75	750	.835	2.410	3501	0
P76	760	.847	2.410	3501	0
P77	770	.858	2.410	3501	0
P78	783	.872	2.377	3453	48

COMPANY: OIL CO.
TICKET NO: H111459
GAUGE NO: 3135

SECOND CLOSURE 47, 10 MINUTE INTERVALS. LAST INTERVAL= 8 MINS.
TOTAL 478 MINUTES

	TIME		PRESSURE		
	MINUTES	INCHES DEFL.	INCHES DEFL.	PSIG	DELTA P
P0	0	.000	2.377	3453	0
P1	10	.011	2.613	3798	+345
P2	20	.022	2.613	3798	0
P3	30	.033	2.612	3796	2
P4	40	.045	2.612	3796	0
P5	50	.056	2.611	3795	-1
P6	60	.067	2.611	3795	0
P7	70	.078	2.610	3793	-2
P8	80	.089	2.610	3793	0
P9	90	.100	2.610	3793	0
P10	100	.111	2.610	3793	0
P11	110	.123	2.610	3793	0
P12	120	.134	2.610	3793	0
P13	130	.145	2.610	3793	0
P14	140	.156	2.610	3793	0
P15	150	.167	2.609	3792	-1
P16	160	.178	2.608	3790	-2
P17	170	.189	2.608	3790	0
P18	180	.201	2.608	3790	0
P19	190	.212	2.608	3790	0
P20	200	.223	2.608	3790	0
P21	210	.234	2.608	3790	0
P22	220	.245	2.608	3790	0
P23	230	.256	2.608	3790	0
P24	240	.267	2.608	3790	0
P25	250	.278	2.608	3790	0
P26	260	.290	2.608	3790	0
P27	270	.301	2.608	3790	0
P28	280	.312	2.608	3790	0
P29	290	.323	2.608	3790	0
P30	300	.334	2.608	3790	0
P31	310	.345	2.608	3790	0
P32	320	.356	2.608	3790	0
P33	330	.368	2.608	3790	0
P34	340	.379	2.608	3790	0
P35	350	.390	2.608	3790	0
P36	360	.401	2.608	3790	0
P37	370	.412	2.608	3790	0
P38	380	.423	2.608	3790	0
P39	390	.434	2.608	3790	0
P40	400	.446	2.608	3790	0
P41	410	.457	2.608	3790	0
P42	420	.468	2.608	3790	0
P43	430	.479	2.608	3790	0
P44	440	.490	2.608	3790	0
P45	450	.501	2.608	3790	0
P46	460	.512	2.608	3790	0
P47	470	.524	NO READINGS POSSIBLE		
P48	478	.533			

DST SUMMARY

Company____________________ Date of Test________

Well__________ DST No.__________ Zone__________

Field__________ Area__________ Country__________

Elevation ______Ft. Test Tool__________

Total Depth ______Ft. ________Depth________Ft.

Producing Interval ______Ft. Downhole Choke __________

Casing Size______Depth______Ft. Surface Choke __________

Liner Size ______Depth______Ft. Separator Choke __________Ft.

Tubing/Drill Pipe Size ______Ft. Fluid Cushion __________Ft.

Total Flow Period ______Ft. From: ________ to: ________

Stable Flow Period ______Ft. From: ________ to: ________

PRODUCTION

BFPD________ BOPD________ BCPD________ BWPD________

BS&W________ Salinity________ RW=________ohm-meters @______°F

Oil Gravity @ 60°F ________°API

Total Gas ________MCF/Day

First Stage ________MCF/Day Specific Gravity________

Second Stage ________MCF/Day Specific Gravity________

Gas Oil Ratio ________SCF/Bbl Stock Tank Oil ________

THP (Flowing)________ THP (Shut In)________ Build up Time________

RESERVOIR PARAMETERS

ISIP________ IFP________ BHT (Max Temp Thermometer)________°F

FFP________ FSIP________ IHP________ FHP________

Type Recorder____________________

Productivity Index____________________ (Bopd/Psi Drawdown)

REMARKS:

Appendix C

The following represents a sample calculation of a drillstem test.

Well #2

DST #6

L. Jurassic Limestone
(same section as DST#5)

Drawdown: Shaky analysis possible from plot of Log Δt vs P_{wf}.

Data: $m = 106$ psi/cycle
$q_o = 6006$ STB/day
$\mu_o = 1.16$ cp
$B_o = 1.17$ B/STB

} (from fluid analysis Avg. @ P= 3675 psig)

$$kh = \frac{162.6\, q\mu B}{m} = \frac{(162.6)(6006)(1.16)(1.17)}{106} = \boxed{12{,}504 \text{ md. ft}}$$

$h = 64\,m = 210$ ft (log analysis)

$$K_{eff} = \frac{12{,}504}{210} = \boxed{59.5 \text{ md.}}$$

Data: $S_w = 22.1\%$ (log analysis)
$\phi = 8.6\%$ (log analysis)
$\phi_{HC} = (0.086)(1-0.221) = 0.067$ or 6.7%
$P_i = 3830$ psig
$P_{1Hr} = 3653$ psig
$r_w = 0.25$ ft (7" liner)
$c = 8.97 \times 10^{-6}$ vol/vol/psi (fluid data)

$$S = 1.15\left[\frac{P_i - P_{1Hr}}{m} - \log\left(\frac{k}{\phi\mu c r_w^2}\right) + 3.23\right]$$

$$s = 1.15\left[\left(\frac{3830-3653}{106}\right) - \log\left(\frac{(59.5)(10^6)}{(0.067)(1.16)(8.97)(0.25)^2}\right) + 3.23\right]$$

$$s = (1.15)\left[1.670 - \log\left(1.366 \times 10^9\right) + 3.23\right]$$

$$s = \boxed{4.87}$$

No Build-up analysis possible from this test.

CS-2 Well
DST #6
2nd Flow

U. Jurassic

Time	Δt				DST Chart Δt	DST Chart Δt
2218 →	0		16/64		-29	-16
2300 →	42	choke change	32/64		13	26
2305 →	47		64/64		18	31
2312 →	54	positive bean	64/64		25	38
2324 →	66	choke change	32/64		37	50
2343 →	85	positive bean	32/64		56	69
0149 →	211	choke change	80/64		182	195
0215 →	237	(1½" choke adj.)	80/64		208	221
0256 →	278	choke change	112/64	(Gradual Inc	249	262
0300 →	282	(1¾" choke on adj.)	112/64	128/64	253	266
0328 →	310	(choke fully opened)	128/64 ←		281	294
0340 →	322	positive bean	128/64		293	306
0400 →	342	commence flowrate at sep	128/64		313	326
1130 →	792	last P & Q data	128/64		763	776
1137 →	799	tester closed for B.U.	—		770	783

799 here ~ (770-783) on DST chart

Seems to match better with choke settings and BHP chart readings

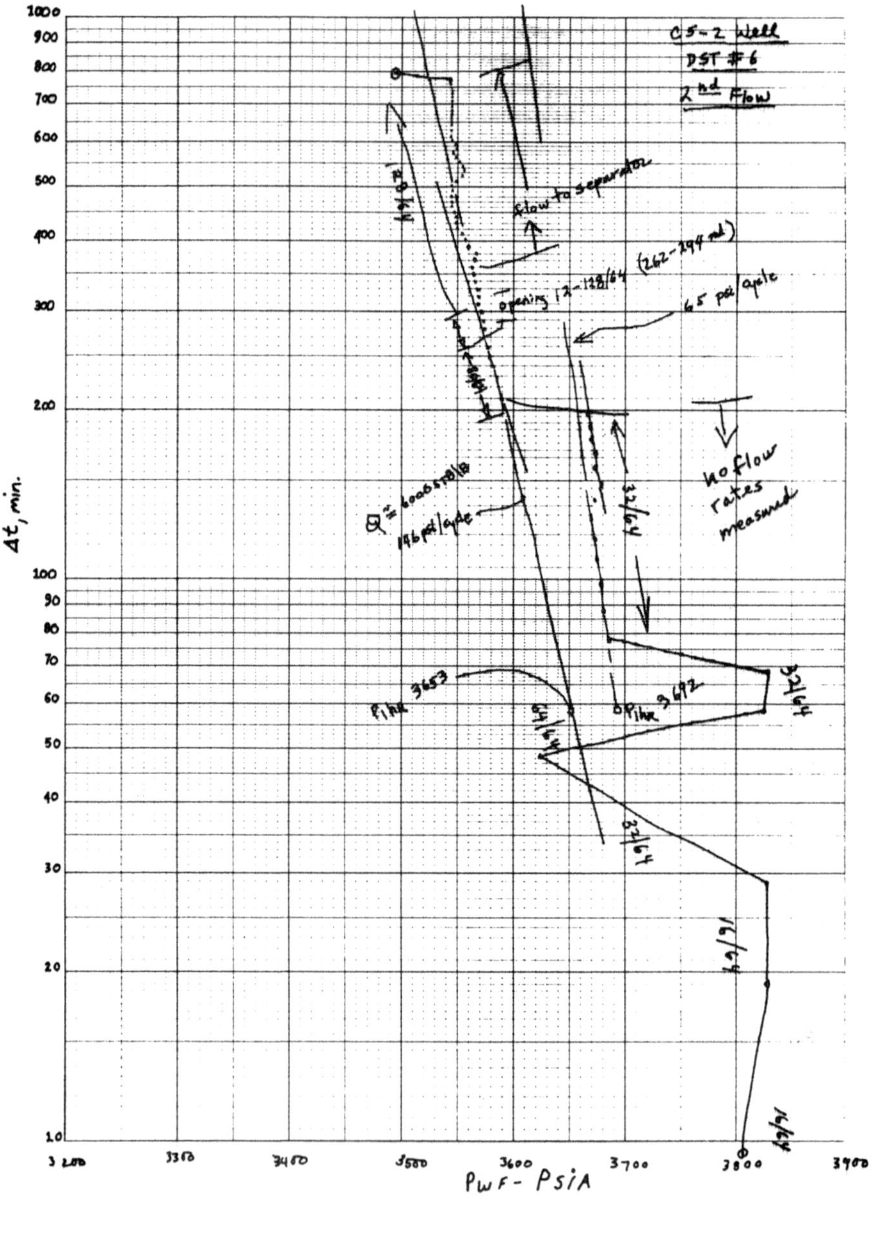

C5-2 Well
DST #6
2nd Flow
flow to separator
opening 12-128/64 (262-294 md)
65 psi/cycle
146 psi/cycle
no flow rates measured
32/64
64/64
16/64
P1hr 3653
P1hr 3692
Δt, min.
PWF - PSIA
1000
900
800
700
600
500
400
300
200
100
90
80
70
60
50
40
30
20
10
3200
3300
3400
3500
3600
3700
3800
3900

Index

About the Author

Ellis H. Austin is currently working as a consultant to the petroleum industry. Recent projects have included enhanced oil recovery of steam, gas, and improved water flood, as well as geopressured, geothermal drilling. His more than thirty-eight years of experience in various phases of the petroleum and related industries include drilling operations in the Gulf of Suez for Transworld Egypt Petroleum Group, facilities design in the Argyll Field for Hamilton Brothers, Production Superintendent on N.W. Offshore Java for Atlantic Richfield Indonesia, and Maintenance Supervisor for Occidental of Libya. Additionally, he spent twenty years in oilfield sales and service and has owned and operated an equipment service and maintenance company. He has written a book on production calculations and has written papers on subsurface production chemicals and electric motors. He holds a B.S. degree in petroleum and natural gas engineering and an LL.B. degree.

Short Course Handbooks Series

N. A. Anstey, *Simple Seismics*

F. A. Guiliano (ed.), *Introduction to Oil and Gas Technology,* 2nd edition

M. G. Barbier, *Pulse Coding in Seismology*

R. C. Selley, *Petroleum Geology for Geophysicists and Engineers*

Other books in petroleum engineering:

E. J. Burcik, *Properties of Petroleum Reservoir Fluids*

D. A. T. Donohue/T. Ertekin, *Gaswell Testing: Theory, Practice and Regulation*

F. A. Guiliano (ed.), *Introduction to Oil and Gas Technology,* 2nd edition

G. A. Karim/A. B. Hamilton, *Metrication for the Exploration and Production Professional*

M. Muskat, *The Flow of Homogeneous Fluids Through Porous Media*

M. Muskat, *Physical Principles of Oil Production*

G. W. Thomas, *Principles of Hydrocarbon Reservoir Simulation* 2nd edition

Also:

J. J. Connor (ed.), *Petroleum Industry Training Resource Guide*
3 issues/year, ISSN 0736-6221
Listings of available courses, books, and audio-visual programs compiled for petroleum personnel from suppliers worldwide, arranged by job category and subject.

GPSR Compliance

The European Union's (EU) General Product Safety Regulation (GPSR) is a set of rules that requires consumer products to be safe and our obligations to ensure this.

If you have any concerns about our products, you can contact us on ProductSafety@springernature.com

In case Publisher is established outside the EU, the EU authorized representative is:

Springer Nature Customer Service Center GmbH
Europaplatz 3
69115 Heidelberg, Germany

Batch number: 10614713

Printed by Printforce, the Netherlands